SpringerBriefs in Mathematics

Series Editors

Nicola Bellomo, Torino, Italy

Michele Benzi, Pisa, Italy

Palle Jorgensen, Iowa City, USA

Roderick Melnik, Waterloo, Canada

Otmar Scherzer, Linz, Austria

Benjamin Steinberg, New York, NY, USA

Lothar Reichel, Kent, USA

Yuri Tschinkel, New York, NY, USA

George Yin, Detroit, USA

Ping Zhang, Kalamazoo, MI, USA

David Levanony • Peter E. Caines

Stochastic Lagrangian Adaptation

 Springer

David Levanony (iD)
School of Electrical and Computer
Engineering
Ben Gurion University
Beer Sheva, Israel

Peter E. Caines (iD)
Department of Electrical and Computer
Engineering
McGilll University
Montreal, QC, Canada

ISSN 2191-8198   ISSN 2191-8201   (electronic)
SpringerBriefs in Mathematics
ISBN 978-3-031-73757-2   ISBN 978-3-031-73758-9   (eBook)
https://doi.org/10.1007/978-3-031-73758-9

Mathematics Subject Classification: 60G17, 93E35, 62F12, 62F30 , 93D21, 93E20, 93E24

This Springer imprint is published by the registered company Springer Nature Switzerland AG
The registered company address is: Gewerbestrasse 11, 6330 Cham, Switzerland

If disposing of this product, please recycle the paper.

Contents

1 Introduction 1
   1.1 Methodology and Principal Results 3

2 Problem Statement 7

3 Asymptotic Maximum Likelihood Identification 11
   3.1 Asymptotic Maximum Likelihood Estimates 12
   3.2 The Characterization of the Limit Set of AML Estimates 14
   3.3 An Alternative Definition of the Limit Set $\mathcal{I}$ 24
   3.4 A Class of AML Estimates 26

4 Geometric Results 27
   4.1 A Conflict Between Identification and Control 27
   4.2 The Unique Stationary Point of J over $\mathcal{I}$ 28
   4.3 The Scalar Example 32

5 Lagrangian Adaptation 33
   5.1 Motivation and Sketch of the Conceptual Adaptive Scheme 34
   5.2 An Estimate of the B-Derivative of the Closed-Loop Matrix 36
   5.3 Derivation of the Estimate and the Adjoint Processes Flows 41
   5.4 The Validity of the Lagrangian Adaptive Control Scheme 44
   5.5 From Convergence to Consistency to Optimal Performance 49
   5.6 A Discussion of the Adaptive Scheme's Initialization Time t_0 51
   5.7 A Summary of the Lagrangian Adaptation Scheme 51
   5.8 Computation of the Regret Rate 54

6 Proof of Theorem 5.2 57
   6.1 Proof Methodology 57
   6.2 Preliminary Results 58
   6.3 A Zero-Order Approximation of Π 60

6.4   A First-Order Approximation of Π 61

6.5   Generalized Stochastic Picard Iterations 63

6.6   Convergence of $\{\Pi_t^n, \zeta_t^n \ t \geq 0\}$ to Solutions of (5.10, 5.11) 64

References 73

Index 75

Chapter 1
Introduction

Abstract An overview of stochastic adaptive control from its birth to date is provided. In particular, the concept of Certainty Equivalence (CE) adaptive control is highlighted together with the fact that in some of its principal forms it requires consistent parameter estimation. Since sample-path Persistent Excitation (PE) of the associated regression vector process ensures consistency, various attempts to secure that property are discussed. The PE requirement may be abandoned when certain control-biased estimation methods are utilized. One such methodology is realized in this work which is built upon a geometric study of limit sets of parameter estimates which give rise to closed-loop indistinguishable dynamics.

Keywords Stochastic adaptive control (SAC) · Certainty equivalence (CE) · Recursive least squares (RLS) · Vanishing dither injection · Stochastic differential equation (SDE) · Strong consistency · Persistent excitation (PE) · Constrained optimization · Linear quadratic (LQ) adaptive control · Set of indistinguishable closed-loop dynamics · Biased parameter estimates

One of the earliest formulations of stochastic adaptive control (SAC) was the self-tuning regulator (STR) introduced by Åström and Wittenmark [1]. The STR is a Certainty Equivalence adaptive control scheme for systems described by discrete-time autoregressive moving average (ARMAX) models relating system control inputs, measured outputs and exogenous disturbance processes. Certainty Equivalence adaptive controllers consist of: (i) a learning mechanism which recursively generates parameter estimates, and (ii) a standard control law with the values of the unknown system parameters replaced by their estimates at any instant in time. The STR of Åström and Wittenmark [1] and Clarke and Gawthrop [12] generates recursive least squares (RLS) parameter estimates which are used in an output feedback minimum variance control law. Properties of the behavior of the STR scheme were obtained in that work subject to the assumption of the consistency (i.e., convergence to true values) of the parameter estimates.

Subsequently, in the work of Goodwin, Ramadge and Caines [19] (see Goodwin and Sin [18], and Caines [4, Chapter 12]) the first complete asymptotically optimal stochastic adaptive control result was established. This was for the stochastic gradient parameter estimation algorithm in a Certainty Equivalence adaptive controller configuration; it was shown that this yielded almost sure asymptotic (long-term average) minimum variance tracking of a bounded deterministic sequence. The conditions imposed on the system are those of inverse stability (which is necessary for minimum variance control) and a positive real property on the spectrum of the disturbance process. The system is not required to be stable. Subsequently, Goodwin and Sin (see [18]) and Chen and Guo [11] extended these results to the Modified Recursive Least Squares (MRLS) algorithm (see, e.g., [4, Chapter 12]).

Consistency of parameters estimates is not a requirement for the asymptotically optimal adaptive control results above to hold nor is it a consequence (see, e.g., Becker et al. [2]). In fact, for ARMAX systems, asymptotically optimal SAC properties hold for parameter estimates which merely converge into the equivalence class of parameters of optimal input–output controllers (see [4, Chapter 12]). However consistency was ensured in some SAC methods via a condition on the comparative rates of growth of the eigenvalues of the empirical covariance of the system regression vector (see, e.g., Lai and Wei [28, 29]). For the adaptive control scheme of Goodwin et al. consistency with an arbitrarily small loss of optimality was first obtained by the injection of stationary dither signals by Caines and Lafortune [6]. For MRLS, Chen and Guo [11] subsequently established asymptotic optimality with strong consistency via the design of an asymptotically vanishing dither injection.

Stochastic adaptive control in the continuous-time parameter case is naturally formulated in terms of stochastic differential equations (SDEs) with states evolving in finite-dimensional space and that is the framework employed in this monograph. In the setting of continuous-time parameter SAC, strong consistency is invariably a consideration for the system parameter estimates which are used to recursively generate state feedback controllers. Indeed, such consistency properties were established in the work of Duncan et al. [16] in the derivation of asymptotic (long-range average) optimality SAC results.

The work of Kumar and Becker [23] and Kumar and Lin [24] introduced the technique of biasing standard parameter estimation schemes in adaptive control algorithms for controlled Markov chains. This biasing is a function of the system performance that would result if the true system were described by the current parameter estimate. The basic idea is to direct the process of parameter estimates in a way that takes into account of both identification and steady-state system performance. This line of research was continued for controlled diffusion processes by Borkar [3], Campi and Kumar [10], and Prandini and Campi [37]. One notable advantage of this approach is that it permits a significant weakening of the persistent excitation (PE) requirement that is to be found in the work on consistency-based stochastic adaptive control in the continuous-time case (see, e.g., Dunan and Pasik-Duncan [13–15], Duncan et al. [16], Caines [5]).

The adaptive control scheme proposed in this work involves parameter estimates that are implicitly biased. Unlike the work cited above, these estimates are a product of the solution of a time-varying constrained optimization problem, which is constructed so as to make its solution inherently consistent. The use of these biased estimates within a certainty equivalence control scheme results in asymptotically optimal LQ performance.

One contribution of this monograph lies in the fact that the adaptive control algorithm described is completely recursive, in contrast to that of Borkar [3], which implicitly involves an instantaneous optimization procedure.

A particular feature of our approach is the use of the geometric analysis of Polderman [34–36] who studied the structure of the sets of parameters corresponding to (i) indistinguishable closed-loop dynamical behavior and (ii) optimal LQ closed-loop performance. This analysis was specifically carried out with the study of LQ adaptive control schemes in view.

The authors' recent work [8] proposed an algorithm which involves alternating gain matrices in a control policy that is shown to guarantee consistent parameter estimation. This comes at the cost of an inherent ε-optimality of the resulting control performance. In addition, that scheme entails a parameter estimate resetting mechanism which adds to the complexity of its utilization. The adaptive scheme presented in this monograph is a refinement of the scheme outlined in Levanony and Caines [32]. In effect, it trades the aforementioned complexity and ε-optimality for algorithmic complexity of a different form so as to result in an adaptive scheme providing optimal closed-loop performance.

The recent literature (see, e.g., [17, 39] and references therein) on stochastic adaptive linear quadratic regulation establishes consistent estimation together with low regret ($\mathcal{O}(1/\sqrt{T})$) with high probability. However, this work shares the common assumption that there is a known stabilizing feedback system gain or simply that the systems are open loop stable. This is in contrast to the situation considered in the present analysis and its predecessors [7, 8], where, without any a priori stability assumptions, adaptive stabilization is established on almost all system sample paths.

It is to be noted that our exposition is founded on the basic results derived in [8, Sections 3 and 4] (*Asymptotic Maximum Likelihood Identification* and *Geometric Results*, respectively). Therefore, for completeness, following the arguments given in [8], we reiterate these results, together with their proofs, in Chaps. 3 and 4.

1.1 Methodology and Principal Results

Consider a class of completely observed LTI systems whose states evolve according to the Îto equations

$$dx_t = Ax_t dt + Bu_t dt + C dw_t,$$

where x, u, w take values in $\mathbb{R}^n, \mathbb{R}^m, \mathbb{R}^r$ (respectively) and w is a standard Brownian process independent of x. As in all related publications cited in this work, full state observations are considered. For a given system, parameterized by unknown $(A^*, B^*) \triangleq \theta^* \in \mathcal{S}$, $\mathcal{S}$ being the set of all stabilizable pairs (A, B), our objective is to recursively generate estimates $\{\theta_t; t \geq t_0\}$ to be used in an LQ certainty equivalence control to obtain the optimal LQ (long run) cost which would be obtained if θ^* were known. In the solution to this problem provided in this monograph, we consider a certainty equivalence based control policy as described in Caines [5]. The basic difference lies in the parameter estimation algorithm. In [5] a standard RLS algorithm has been employed under a (sample-wise) PE condition which requires independent verification. In the adaptive parameter estimation and control work of Duncan and Pasik-Duncan [13, 14], the role of this PE condition is effectively replaced by the condition that a certain determinant $det \widetilde{A}_t$ shall be almost surely bounded away from zero. This condition is verifiable in certain cases of interest. Alternatively, the required PE property may be created by the injection of a diminishing excitation signal, as has been shown by Duncan, Guo and Pasik-Duncan [16]. In the present work, no external dither is utilized, nor any sample-wise PE property is assumed to hold.

Let $\varphi_t^T = (x_t^T, u_t^T) = (x_t^T, -x_t^T K_t^T)$ denote the regression vector where $K_t = K(\theta_t)$ is computed via the control law; then, roughly speaking, one class of PE conditions is equivalent to assuming that the matrix $\int_0^t \varphi_s \varphi_s^T ds$ (properly normalized) converges to a strictly positive-definite limit. As mentioned earlier, another important class is that involving the comparative rates of growth of eigenvalues of this matrix (see Lai and Wei [28, 29], Chen and Guo [11]). Note, however, that if $K_t \rightarrow K$, the existence of such a positive-definite limit is questionable, as the state x and the control $u = -K(\theta)x$ become asymptotically linearly dependent. PE conditions are almost invariably used to ensure consistency; an alternative formulation, which we pursue in this work, is that where one simply considers the natural convergence $\theta_t \rightarrow \theta_\infty \in \mathcal{I}$, as $t \rightarrow \infty$, where $\mathcal{I}$ is some limit set, which occurs without further conditions for essentially all parameter estimation algorithms. (In our case, $\mathcal{I}$ corresponds to parameters yielding indistinguishable system trajectories.)

Let $J(\theta)$ be the long-run optimal performance cost for a system with $(A, B) = \theta$. As has been established by Polderman [34–36], the limit set of indistinguishable dynamics

$$\mathcal{I} = \{(A, B) | A - BK(A, B) = A^* - B^*K(A, B)\}$$

is a smooth manifold upon which J has a unique minimum at θ^* (= the true parameter) and which has no local stationary points other than θ^* on $\mathcal{I}$. With these properties in mind, the intuition behind the performance biased adaptive control algorithm is as follows: Suppose one uses the conventional certainty equivalence adaptive controller, combined with an ML-type parameter estimation algorithm. Such a scheme uses current parameter estimates to generate approximately optimal (in a certain sense) predictions of the state trajectories, together with a feedback control law. This has the standard implicit effect that prediction errors are used to adjust the parameter estimates so that, asymptotically, the estimated and the actual closed-loop system dynamics will match exactly. By the discussion above, estimated and actual system dynamics for the closed-loop system will match exactly on the set $\mathcal{I}$, but performance will be suboptimal for any $\theta \in \mathcal{I}, \theta \neq \theta^*$. Therefore, the following refinement is introduced: A performance biased cost is embedded within a constrained optimization problem so that its recursive solution constitutes the parameter estimation part of the certainty equivalence based adaptive scheme. As a result, the overall algorithm eventually drives the parameter estimates in the direction of $-\nabla^{\mathcal{I}} J(\theta)$ (a projection of ∇J on $\mathcal{I}$—see Chap. 4), while due to the specific time-varying constraint it maintains its drive toward $\mathcal{I}$. This, together with the unique minimum property of J on $\mathcal{I}$, results in the desired consistency which in turn, implies the optimality of the long-run performance.

Our results are organized as follows: The adaptive control problem is formulated in Chap. 2. In Chap. 3 a class of AML (Asymptotic Maximum Likelihood) estimates is defined and is shown to have a common limit set $\mathcal{I}$. Chapter 4 is devoted to the investigation of the geometric properties of the limit set $\mathcal{I}$. In Chap. 5, after a formal construction of the algorithm, summarized by equations (5.21–5.25) and (5.41–5.43) below, the following are established: (1) The SDE defined by the proposed algorithm has a unique strong solution; (2) There exists a $\theta_\infty(\omega)$ such that the parameter estimates $\theta_t \to \theta_\infty$, $a.s.$, as $t \to \infty$.; (3) $\theta_\infty \in \mathcal{I}$, $a.s.$; (4) $\theta_\infty = \theta^*$ $a.s.$, and, consequently, optimal, long-run performance cost is obtained (Theorem 5.8).

We observe that in this monograph we follow the often used convention that the Brownian noise intensity matrix C is a priori known. This rests on the fact that, in principle, C can be accurately computed from the quadratic variation of the observed state process $\{x_t, \ t \geq 0\}$, measured over any given small interval $[0, t_0]$.

Chapter 2
Problem Statement

Abstract A solution is presented to the asymptotic minimization problem for the long-run quadratic costs of the controlled state of a linear stochastic differential equation (SDE) system with unknown system parameters and full state observations.

Keywords Stabilizable pair · Algebraic Riccati Equation (ARE) · Asymptotically optimal performance · Maximum likelihood (ML)

We consider the system

$$dx_t = Ax_t dt + Bu_t dt + Cdw_t, \tag{2.1}$$

where $x_t \in \mathbb{R}^n, u_t \in \mathbb{R}^m$ and $w_t \in \mathbb{R}^p$ for all $t \geq 0$, x_0 is a nonrandom initial condition and w is a standard Brownian motion. Let the process $w = \{w_t, \ t \geq 0\}$ be measurable with respect to the increasing family of σ-fields $\mathcal{F}_t$ for all $t \geq 0$. The solution process $x = \{x_t, \ t \geq 0\}$ for (2.1) is dependent upon the values taken by A, B, C, where these are nonrandom, time-independent variables.

Let x generate the increasing family of σ-fields $\mathcal{F}_t^x, t \geq 0$. Then the process $u = \{u_t, \ t \geq 0\}$ is assumed to satisfy the (adaptive) non-anticipative control condition that u_t is only an $\mathcal{F}_t^x$ measurable function for $t \geq 0$ and hence is not an explicit function of A, B, C. We shall denote this condition by $u \in \mathcal{U}$.

The objective in the application of adaptive control to the system (2.1) is to achieve, along almost all sample paths,

$$J_\infty^o \triangleq \inf_{u \in \mathcal{U}} \lim_{t \to \infty} \frac{1}{t} \int_0^t (\|x_s\|^2 + \|u_s\|^2) ds, \tag{2.2}$$

where the control laws employed are such that the inner limit exists, along almost all paths.

Let us define the matrix parameter $\Theta = [A, B]^T$ and the corresponding vector parameter $\theta = [col(A, B)] \in \mathbb{R}^{n(n+m)}$, where $[col(A, B)]^T = [A_1, A_2 \cdots A_n, B_1, B_2 \cdots B_n]$, with A_i (respectively, B_i) denoting the i-th row

D. Levanony, P. E. Caines, *Stochastic Lagrangian Adaptation*, SpringerBriefs in Mathematics, https://doi.org/10.1007/978-3-031-73758-9_2

of A (respectively B). Further define the $n \times (n + m)n$ matrix Ψ_t by

$$\Psi_t = \begin{bmatrix} \varphi_t^T & 0 & \cdots & 0 \\ 0 & \varphi_t^T & 0 & 0 \\ 0 & \cdots & \cdots & \varphi_t^T \end{bmatrix}, \tag{2.3}$$

where $\varphi_t^T \doteq (x_t^T, u_t^T)$. Assuming a full rank noise (i.e., $CC^T > 0$), we take for simplicity $C = I$. The system equation (2.1) may be conveniently reexpressed as

$$dx_t = \Psi_t \theta dt + dw_t, \quad t \geq 0. \tag{2.4}$$

For convenience, when $u \in \mathcal{U}$, we shall refer to both (2.1) and (2.4) as the system $\Xi(\theta)$.

We use the notation θ^* to denote the value of the deterministic parameter of the system (2.4) generating a given set of observations for a given control law $u \in \mathcal{U}$. This parameter θ^* will be referred to as the true system parameter, and we assume that

$$\theta^* \in \mathcal{S} \triangleq \{\theta = col(A, B) : (A, B) \text{ stabilizable}\}.$$

The adaptive control algorithms studied in this monograph are based upon the class of certainty equivalence (CE) algorithms which have the following form for the adaptive LQ problem: For each $t \geq 0$:

 (i) Compute an estimate $\theta_t \in \mathcal{S}$ of $\theta^* \in \mathcal{S}$.
 (ii) Use the feedback control law.

$$u_t = -K(\theta_t)x_t,$$

where

$$K(\theta) = K(A, B) = B^T V(A, B) \tag{2.5}$$

with $V(A, B)$ being the (unique) positive-definite symmetric solution to the Algebraic Riccati Equation (ARE):

$$A^T V + VA - VBB^T V + I = 0. \tag{2.6}$$

It is well known (e.g., Duncan and Pasik-Duncan [13]) that for any given system, parameterized by a stabilizable pair (A, B), the optimal achievable performance (2.2) equals

$$J(A, B) \triangleq \mathrm{Tr} V(A, B). \tag{2.7}$$

Therefore, (with a slight abuse of notation) we shall refer to $J(\theta) \overset{\Delta}{=} \mathrm{Tr} V(\theta)$ as the (synthetic) cost function. As will be shown below, a minimization of (2.7), which takes place during the adaptation procedure, leads to asymptotically optimal performance in the sense that, along almost all sample paths, one obtains

$$\lim_{t \to \infty} \frac{1}{t} \int_0^t (\|x_s\|^2 + \|u_s\|^2)ds = J(\theta^*) = J_\infty^o. \tag{2.8}$$

One of the motives for the use of the gradient search maximum likelihood (ML) class of parameter estimation schemes in the adaptive control algorithm introduced in this monograph is that such schemes are sufficiently flexible to permit various modifications to the algorithm while retaining consistency; another motive is that they are comparatively efficient numerically.

Chapter 3
Asymptotic Maximum Likelihood Identification

Abstract A family of asymptotic maximum likelihood (AML) estimates is defined within the completely observed state process framework and is shown to possess the following strong properties: (i) AML estimates converge to finite limits; (ii) AML limits lie in specified limit sets of parameters giving rise to indistinguishable closed-loop dynamics; (iii) a particular family of AML estimates is shown to drive the associated log-likelihood gradient to zero exponentially fast.

Keywords Asymptotic maximum likelihood (AML) · Bayesian embedding (BE) · Recursive least squares (RLS) · Limit set · Convergence of the closed-loop dynamics parameter · Auxiliary RLS estimate

In this section we recall the results of Caines and Levanony [8, Section 3] where a family of Asymptotic Maximum Likelihood (AML) estimates is defined and studied. Specifically, [8, Section 3] key contributions are: It is shown that any AML estimate is characterized with (i) asymptotically agreeing with the (standard) Maximum Likelihood (ML) estimate, (ii) has a finite limit θ_∞ [8, Lemma 3.1], which, moreover, (iii) belongs to the limit set $\mathcal{I}$ being the set of all parameters giving rise to associated closed-loop dynamics that match the actual, asymptotic closed-loop dynamics [8, Theorem 3.4].

While (i) is a direct consequence of the ML and AML estimates relation, (ii) and (iii) are shown to hold using a three-layered approach: (1) First, by looking at an auxiliary identification problem of a Gaussian unknown $\theta^*(\omega)$ and its associated Bayesian estimate $\widehat{\theta}_t = \mathrm{E}[\theta^*|\mathcal{F}_t^x]$, (ii) and (iii) are established for $\widehat{\theta}_t$ [8, Theorem 3.2]. (2) The second step constitutes the utilization of the Bayesian Embedding (BE) approach. Returning to the original identification problem in which θ^*, the parameter generating the observation, is a nonrandom unknown, one exploits the duality between Kalman filter estimation of a Gaussian parameter and Recursive Least Squares (RLS) estimation of an unknown nonrandom parameter. From this, one concludes that (ii) and (iii) hold for θ_t^{RLS} (the RLS estimate of the original nonrandom parameter) a.e. outside a Lebesgue null set in $\mathbb{R}^{n(n+m)}$ [8, Corollary 3.3]. (3) The last step is then constructed so as to show that the results (ii) and (iii) above hold for any AML estimate. This is done by explicitly computing θ_t in terms of θ_t^{RLS} [8, Theorem 3.4].

A final remark on the basic properties which lead to the above results is the following: Property (ii), that $\theta_t \rightarrow \theta_\infty$, rests on the AML property that $\nabla L_t(\theta_t) \rightarrow 0$, with $\nabla L_t(\theta)$ being the log-likelihood gradient at θ.

3.1 Asymptotic Maximum Likelihood Estimates

Let $\{(u_t, x_t), \ t \geq 0\}$ denote an observed input-state sample path of the system $\Xi(\theta^*)$, and let $L_t(\theta) \triangleq L_t(\theta, (x_0^t, u_0^t)) = \int_0^t \theta^T \Psi_s^T dx_s - \frac{1}{2} \int_0^t \|\Psi_s \theta\|^2 ds$ denote a log-likelihood function of (u_0^t, x_0^t) at any $\theta \varepsilon S$ whose gradient $\nabla L_t(\theta)$ is given by

$$
\begin{aligned}
\nabla L_t(\theta) &= \int_0^t \Psi_s^T dx_s - \int_0^t \Psi_s^T \Psi_s ds \theta \\
&= \int_0^t \Psi_s^T dw_s - \int_0^t \Psi_s^T \Psi_s ds (\theta - \theta^*) \triangleq m_t - \Phi_t \tilde{\theta},
\end{aligned}
\tag{3.1}
$$

where $m_t, t \geq 0$ is an $n(n+m)$ dimensional martingale, $\tilde{\theta} \triangleq \theta - \theta^*$ and

$$
\Phi_t \triangleq \int_0^t \Psi_s^T \Psi_s ds.
\tag{3.2}
$$

Define an ML estimate θ_t^{ML}, as a parameter trajectory which satisfies $\nabla L_t(\theta_t^{ML}) = 0$ for all $t \geq 0$, and, further, an AML estimate, θ_t, to be a process which satisfies

$$
\nabla L_t(\theta_t) \rightarrow 0, \ as \ t \rightarrow \infty, \ a.s.
\tag{3.3}
$$

Note that by (3.1), for all sufficiently large t such that $\Phi_t > 0$,

$$
\theta_t^{ML} = \Phi_t^{-1} \int_0^t \Psi_s^T dx_s,
\tag{3.4}
$$

and, with (3.1), one has that any parameter process θ_t is given by

$$
\theta_t = \Phi_t^{-1} \left(\int_0^t \Psi_s^T dx_s - \nabla L_t(\theta_t) \right).
\tag{3.5}
$$

While (3.5) indeed describes *any* $\mathcal{F}_t^x$ adapted $\theta_t \varepsilon \mathbb{R}^{n(n+m)}$, property (3.3) makes it an *AML* estimate (labeled below as $\theta_t \varepsilon \mathcal{AML}$), asymptotically agreeing with the ML estimate.

In terms of (3.1) we may present the following preliminary maximum likelihood estimation (MLE) result, which we note does not constitute a consistency result without the addition of further hypotheses.

Lemma 3.1 ([8, Lemma 3.1]) *For the system* $\Xi(\theta^*)$, *and the observed process* $\{(u_t, x_t), \ t \geq 0\}$, *let* $\{\theta_t = \theta_t(\theta^*, \omega), \ t \geq 0\}$, *be a process which is progressively measurable with respect to the* σ-*fields* $\mathcal{F}_t^x, t \geq 0$. *Assume that* $\{\theta_t, \ t \geq 0\}$ *belongs to the class of asymptotic ML estimates* ($\mathcal{AML}$) *in the sense that*

$$\nabla L_t(\theta_t) \to 0 \qquad \text{a.s. as } t \to \infty, \tag{3.6}$$

which is denoted by $\{\theta_t, \ t \geq 0\}\varepsilon\mathcal{AML}$. *Further assume that almost surely* Φ_t *is non-singular for all t sufficiently large. Then there exists an a.s. finite random variable* $\theta_\infty = \theta_\infty(\theta^*, \omega)$ *for which*

$$\theta_t \to \theta_\infty \qquad \text{a.s. as } t \to \infty \tag{3.7}$$

for all $\theta^* \notin \mathcal{N}$, *where* $\mathcal{N}$ *is a Lebesgue null set in* $\mathbb{R}^{n(n+m)}$ *independent of* ω.

Proof The proof is based on the convergence of a related recursive least squares (RLS) estimate defined via

$$d\theta_t^{RLS} = R_t^{-1}\Psi_t^T[dx_t - \Psi_t\theta_t^{RLS}dt], \tag{3.8}$$

$$dR_t = \Psi_t^T\Psi_t dt, \quad t \geq 0, \tag{3.9}$$

with the initial conditions θ_0^{RLS} and $R_0 > 0$, where the latter immediately gives $R_t = R_0 + \Phi_t, t \geq 0$.

Set $\widetilde{\theta}_t^{RLS} = \theta_t^{RLS} - \theta^*$. Then, with $m_t = \int_0^t \Psi_s^T dw_s$ (see (3.1)), $\widetilde{\theta}_t^{RLS}$ is given by

$$\widetilde{\theta}_t^{RLS} = R_t^{-1}[m_t + R_0\widetilde{\theta}_0^{RLS}]$$

$$= [R_0 + \Phi_t]^{-1}[m_t + R_0\widetilde{\theta}_0^{RLS}]. \tag{3.10}$$

On the other hand, from (3.1), an estimation error $\widetilde{\theta}_t = \theta_t - \theta^*$, for any $\mathcal{AML}$ estimate θ_t, can be written as

$$\widetilde{\theta}_t = \Phi_t^{-1}[m_t - \nabla L_t(\theta_t)] \tag{3.11}$$

(for all sufficiently large t's), and so eliminating m_t between (3.10) and (3.11) yields

$$\widetilde{\theta}_t = [I + \Phi_t^{-1}R_0]\widetilde{\theta}_t^{RLS} - \Phi_t^{-1}[R_0\widetilde{\theta}_0^{RLS} + \nabla L_t(\theta_t)]. \tag{3.12}$$

By a result of Kumar [26, Theorem 1], there exists a Lebesgue null set $\mathcal{N} \subset \mathbb{R}^{n \times (n+m)}$, independent of ω, such that for any $\theta^* \notin \mathcal{N}, \theta_t^{RLS} \to \theta_\infty^{RLS}(\omega)$, almost surely as $t \to \infty$, where $\|\theta_\infty^{RLS}(\omega)\|$ is a.s. finite.

Assume that $\theta^* \notin \mathcal{N}$; then, since $\nabla L_t(\theta_t) \to 0$ a.s. as $t \to \infty$ by assumption and Φ_t^{-1} is a positive matrix monotonically decreasing with respect to t, we conclude that an $\mathcal{AML}$ estimate $\theta_t, t > 0$, converges a.s. to a limit of finite norm. $\qquad\square$

We observe that the estimates θ_t and θ_t^{RLS} in the theorem above are functions of a given sample path of the $\{(u_t, x_t), t \geq 0\}$ process. In particular, u_t could be a function of the estimate θ_t, and, subject to the hypotheses of the theorem, the result is valid. The introduction of θ_t^{RLS} is purely an auxiliary device to facilitate the proof.

3.2 The Characterization of the Limit Set of AML Estimates

The characterization of the limit set of $\{\theta_t\}\varepsilon\mathcal{AML}$ is the purpose of the final phase of this section. Following the Bayesian embedding approach (Kumar [26]), we begin with a *closed-loop* consistency proof in a Gaussian setting:

Theorem 3.2 ([8, Theorem 3.2]) *Consider the system (2.1) where it is assumed that $C = I$, (A^*, B^*, x_0) have a joint Gaussian distribution and $w = \{w_t, t \geq 0\}$ is a standard Brownian vector process, independent of (A^*, B^*, x_0).*

Suppose that the system is controlled by $u_t = -K_t x_t$, $t \geq 0$, where K_t is a causal, $\mathcal{F}_t^x$-measurable, continuous, and bounded semimartingale (matrix) which converges a.s. to a finite, possibly random limit K_∞. Let

$$\widehat{\Theta}_t \triangleq \mathrm{E}[\Theta^* | \mathcal{F}_t^x] = \mathrm{E}([A^*, B^*]^T | \mathcal{F}_t^x) \tag{3.13}$$

be the MV (minimum variance) estimate of $\Theta^* \triangleq [A^*, B^*]^T$. Then

$$\widehat{\Theta}_t \to \{[A, B]^T : A - BK_\infty = A^* - B^* K_\infty\} \quad a.s. \ as \ t \to \infty. \tag{3.14}$$

Corollary 3.3 ([8, Corollary 3.3]) *Let Θ_t^{RLS} be the matrix RLS estimate corresponding to θ_t^{RLS} of Lemma 3.1 (i.e., $\theta_t^{RLS} = col\{\Theta_t^{RLS}\}$). Then, under the conditions of Theorem 3.2 but with $[A^*, B^*]$ deterministic and stabilizable, Θ_t^{RLS} is a.s. convergent to some Θ_∞^{RLS} with*

$$\Theta_\infty^{RLS} \varepsilon \{[A, B]^T : A - BK_\infty = A^* - B^* K_\infty\}, \ a.s.$$

for all $[A^*, B^*]^T = \Theta^* \notin \mathcal{N}$ where $\mathcal{N}$ is a Lebesgue null set in $\mathbb{R}^{(n+m)\times n}$ independent of ω.

Proof The RLS algorithm (see proof of Lemma 3.1) is, in the Bayesian setup of Theorem 3.2, the Kalman filter for $\mathrm{E}[\Theta^* | \mathcal{F}_t^x)] = \widehat{\Theta}_t$. Hence, the conclusion follows by the convergence result of Theorem 3.2 and the mutual absolute continuity of the Lebesgue and the Gaussian measures on $\mathbb{R}^{(n+m)\times n}$. $\qquad\square$

Proof of Theorem 3.2

A Truncation Mechanism

We first claim that it suffices to prove the theorem for feedback gains $\{K_t, t \geq 0\}$ which make, w.p.1, the resulting state-dynamics matrix process $\{A^* - B^* K_t, \ t \geq 0\}$ uniformly bounded by some nonrandom $N < \infty$. This is based on the following argument: For an a.s. bounded $\{K_t, t \geq 0\}$, let $\tau_N = \inf\{t \geq 0 : \|A^* - B^* K_t\| \geq N\}$ and define an auxiliary process x^N satisfying

$$x_t^N = x_t, \qquad 0 \leq t \leq \tau_N$$
$$dx_t^N = [A^* - B^* K_{\tau_N}] x_t^N dt + dw_t, \quad t \geq \tau_N.$$

Since $\sup_{t \geq 0} \|K_t\| < \infty$ a.s., for each $\varepsilon > 0$, there exists $N_\varepsilon < \infty$ s.t.

$$P(\tau_N = \infty) > 1 - \varepsilon, \quad \forall N \geq N_\varepsilon.$$

Therefore, if we prove that

$$\widehat{\Theta}_t^N \triangleq \mathrm{E}[\Theta^* | \mathcal{F}_t^{x^N}] \to \{[A, B]^T : A - B K_{\tau_N} = A^* - B^* K_{\tau_N}\} \ \text{a.s.}, \forall N < \infty,$$

one may conclude that (3.14) holds, with probability greater than $1 - \varepsilon$. The arbitrariness of ε then yields the proof of the claim.

(Note that the stopped gain matrix process $\{K_t^N = K_t, \ t\varepsilon[0, \tau_n), \ K_t^N = K_{\tau_N}, \ t \geq \tau_n\}$ remains a semimartingale.)

Henceforth, we continue our investigation for the truncated process x^N for which the feedback matrix is uniformly bounded by a deterministic N. In order to avoid cumbersome notations, the superscript N is omitted throughout.

Convergence of the Closed-Loop Dynamics Parameter

Let

$$\alpha_t^* \triangleq col\{[A^* - B^* K_t]^T\} = col\{((I, -K_t^T]\Theta^*)^T\} \tag{3.15}$$

(recall that $\Theta^* = [A^*, B^*]^T$). Since K_t converges a.s.,

$$\alpha_t^* \to \alpha_\infty^* = col\{[A^* - B^* K_\infty]^T\}, \ \text{a.s.} \tag{3.16}$$

With $\widehat{\Theta}_t$ the MV estimate of Θ^*, let

$$\widehat{\alpha}_t \triangleq col\{((I, -K_t^T]\widehat{\Theta}_t)^T\} = \mathrm{E}[\alpha_t^* | \mathcal{F}_t^x] \tag{3.17}$$

be the MV estimate of the closed-loop parameter vector of the system (2.1) under the feedback control $u_t = -K_t x_t$. (Note that the second equality in (3.17) is due to the fact that K_t is $\mathcal{F}_t^x$-measurable.)

Our objective is to show that $\widehat{\alpha}_t$ is strongly consistent in the sense that

$$\widehat{\alpha}_t \to \alpha_\infty^* \qquad \text{a.s.} \qquad \text{as } t \to \infty. \tag{3.18}$$

Define the $n \times n^2$ matrix

$$X_t = \begin{bmatrix} x_t^T & 0 & \cdots 0 \\ 0 & x_t^T 0 & \cdots 0 \\ 0 & \cdots & \cdots x_t^T \end{bmatrix}.$$

Then equation (2.1) is rewritten as

$$dx_t = X_t \alpha_t^* dt + dw_t, \tag{3.19}$$

or equivalently, with $A_t^* = A^* - B^* K_t$ (being the true closed-loop dynamics matrix),

$$dx_t = A_t^* x_t dt + dw_t. \tag{3.20}$$

A Persistent Excitation Property

Next, by the result of [31], it follows from (3.20), together with the fact that $A_t^* \to A_\infty^*$, a.s. (3.16), that the state process $\{x_t\}$ is of *persistent excitation* (PE), in the sense that

$$\liminf_{t \to \infty} \frac{1}{t} \lambda_{\min} \{ \int_0^t x_s x_s^T \, ds \} > 0, \ a.s. \tag{3.21}$$

(where $\lambda_{\min} \{A\} \triangleq$ smallest eigenvalue of A).

Equipped with (3.21), together with the convergence of the closed-loop dynamics parameter process (3.16), the concluding step of the proof will utilize an adaptation of [33, Theorem 1] (using this work's terminology).

A Nonlinear Filtering Problem

Consider the SDE (3.19), taken below as the *measurement equation* when evaluating $\widehat{\alpha}_t = \mathrm{E}[\alpha_t^* | \mathcal{F}_t^x]$, with α_t^* being the $\mathbb{R}^{n^2}$ process defined by (3.15), representing the closed-loop dynamics matrix process.

Since by construction, $\sup_{t \geq 0} \|\alpha_t^*\| \leq N$, (3.16), then Corollary II-2.4 of Revuz and Yor [38] implies that there exists an a.s. finite $\widehat{\alpha}_\infty$ such that

$$\lim_{t \to \infty} \widehat{\alpha}_t = \widehat{\alpha}_\infty = \mathrm{E}[\alpha_\infty^* | \mathcal{F}_\infty^x], \ a.s. \tag{3.22}$$

(where $\mathcal{F}_\infty^x = \sigma(\cup_{t \geq 0} \mathcal{F}_t^x)$). Then, (3.16), (3.22), together with $\widetilde{\alpha}_t = \widehat{\alpha}_t - \alpha_t^*$, $P_t = \mathrm{E}[\widetilde{\alpha}_t \widetilde{\alpha}_t^T | \mathcal{F}_t^*]$, result in

$$\lim_{t \to \infty} P_t = P_\infty = \mathrm{E}[\widetilde{\alpha}_\infty \widetilde{\alpha}_\infty^T | \mathcal{F}_\infty^x], \tag{3.23}$$

with P_∞ being a.s. finite.

Our aim is to show that $\lambda_{max}\{P_\infty\} = 0$ (a.s.), which, together with (3.23), will imply that indeed $\widehat{\alpha}_\infty = \alpha_\infty^*$, a.s.

Write $\alpha_t^* = col\{[A^* - B^* K_t)\} = \alpha_0^* + \int_0^t a_s ds + m_t$, where, respectively, $\{\alpha_0^* + \int_0^t a_s ds, \ t \geq 0\}$ and $\{m_t = \int_0^t D_s dw_s, \ t \geq 0\}$ are the bounded variation (BV) part and the (local) martingale part of $\{\alpha_t^*, \ t \geq 0\}$.

To formalize a *nonlinear* filtering problem, consider the state $\{\alpha_t^*, t \geq 0\}$, evolving according to the implicit nonlinear state equation

$$d\alpha_t^* = d\big(col\{([I, -K_t^T(\theta_t)]\Theta_t^*)^T\}\big) = -col\{B^* dK(\theta_t)\} = a_t dt + D_t dw_t, \tag{3.24}$$

where dK_t is propagated by $d\theta_t$ (itself given by a particular algorithm), via the Ito formula (unspecified further as no need for it arises below).

The state-space model is finalized with the *linear* measurement equation (3.19)

$$dx_t = X_t \alpha_t^* dt + dw_t. \tag{3.25}$$

The innovation process $\{v_t, \ t \geq 0\}$, an $\{\mathcal{F}_t^x\}$ Brownian motion, is written in the form

$$v_t = x_t - \int_0^t X_s \widehat{\alpha}_s ds = \int_0^t X_s \widetilde{\alpha}_s ds + w_t. \tag{3.26}$$

Next, we note that since $\{x_t, \ t \geq 0\}$ is generated by a linear SDE with bounded coefficients, it does not have a finite escape time (a.s.). Specifically, as $\sup_{t \geq 0} \|A_t^*\| < N$, a.s., by virtue of the truncation argument above, it holds that $\widetilde{\Phi}(t, s), \ t \geq s \geq 0$, the transition matrix associated with the state-dynamics matrix process $\{A_t^* = A^* - B^* K_t, \ t \geq 0\}$ is *nonrandomly* bounded in the form,

$$\|\widetilde{\Phi}(t, s)\| \leq e^{N(t-s)}, \ \forall t \geq s \geq 0, \ a.s. \tag{3.27}$$

With (3.27), one may immediately conclude that

$$\int_0^t E[\|x_s\|^2] ds < \infty, \ \forall \ 0 \leq t < \infty. \tag{3.28}$$

A Second Truncation

To further facilitate the use of nonlinear filtering theory, we introduce yet another truncation: Recall that $\{\int_0^t a_s ds, \ t \geq 0\}$ is the BV part of $\{\alpha_t^*, \ t \geq 0\}$. Let

$$T_N = \inf\{t \geq 1 : \int_{t-1}^t |a_s| ds \vee \int_{t-1}^t |\widehat{a}_s| ds > N\}. \tag{3.29}$$

Set $s_N = T_N \wedge \tau_N$ and replace τ_N with s_N, namely, redefine $K_t^N = \{K_t, \ t\varepsilon[0, s_N), \ K_{s_N}, \ t \geq s_N\}$. Then, the (truncated) observed state x_t^N is redefined accordingly, with $A_t^{*N} = A^* - B^* K_t^N$, by

$$x_t^N = x_t, \ 0 \le t \le s_N,$$
$$dx_t^N = A_t^{*N} x_t^N dt + dw_t, \ t > s_N.$$

The truncation argument, stated at the beginning of the proof (see A truncation mechanism), remains valid with s_N replacing τ_N. With the superscript N omitted (unless specifically required), and utilizing Kallianpur [21, Theorem 8.4.3], one may explicitly write

$$\widehat{\alpha}_t = \widehat{\alpha}_0 + \int_0^t \widehat{a}_s ds + \int_0^t [\widehat{D}_s + P_s] X_s^T dv_s, \ t \ge 0, \tag{3.30}$$

where the cross variation of α_t^* and x_t is denoted by $\int_0^t D_s ds = <\alpha^*, x>_t$, with $\widehat{D}_t = \mathrm{E}[D_t | \mathcal{F}_t^x]$.

The Convergence of Increment Processes
Define the collection of unit increment processes,

$$\Delta \alpha^{*t}(s) = \alpha_{t+s}^* - \alpha_t^*, \ s\varepsilon[0, 1], \ t \ge 0. \tag{3.31}$$

Then, by (3.16), it follows that

$$\Delta \alpha^{*t}(\cdot) \to 0, \ \text{in law, as } t \to \infty, \tag{3.32}$$

which, by Jacod and Shiryayev [20, Theorem VI-6.1], gives

$$<\Delta \alpha^{*t}>_1 = \int_t^{t+1} d<m>_s = \int_t^{t+1} D_s D_s^T ds \xrightarrow{P} 0, \ \text{as } t \to \infty. \tag{3.33}$$

Reiterating the same arguments leading to (3.33), now for $\Delta \widehat{\alpha}^t(s) = \widehat{\alpha}_{t+s} - \widehat{\alpha}_t, \ s\varepsilon[0, 1], \ t \ge 0$, one has that, with (3.22) and (3.30), it holds that

$$<\Delta \widehat{\alpha}^t>_1 = \int_t^{t+1} [\widehat{D}_s + P_s] X_s X_s^T [\widehat{D}_s + P_s]^T ds \xrightarrow{P} 0, \ \text{as } t \to \infty, \tag{3.34}$$

hence leading to

$$\liminf_{t \to \infty} \frac{1}{t} \int_0^t [\widehat{D}_s + P_s] X_s X_s^T [\widehat{D}_s + P_s]^T ds = 0, \ \text{in probability.} \tag{3.35}$$

The last result implies the existence of a subsequence $\{t_n\}$ such that

$$\liminf_{n \to \infty} \frac{1}{t_n} \int_0^{t_n} [\widehat{D}_s + P_s] X_s X_s^T [\widehat{D}_s + P_s]^T \, ds = 0, \ a.s. \tag{3.36}$$

Assume that

$$\int_t^{t+1} \widehat{D}_s \widehat{D}_s^T \, ds \xrightarrow{P} 0, \ \text{as } t \to \infty \tag{3.37}$$

(a property shown to hold below).

Recall that $\{m_t = \int_0^t D_s dw_s, \ t \geq 0\}$ is the (local) martingale part of $\{\alpha_t^*, \ t \geq 0\}$. For any large $q < \infty$, consider the stopping time $\tau_q = \inf\{t \geq 0 : \|D_t\| \geq q\}$ and let $D_s^q = D_s \mathbf{1}_{\{s < \tau_q\}}$.

Then, (3.37) obviously gives

$$\int_t^{t+1} \widehat{D}_s^q \widehat{D}_s^{qT} \, ds \xrightarrow{P} 0, \ \text{as } t \to \infty. \tag{3.38}$$

Define the $\mathcal{C}[0, 1]$ processes $\{\widehat{D}_t^q(r) = \widehat{D}_{t+r}^q, \ 0 \leq r \leq 1\}$. Then, given the continuity and boundedness of $D_t^q(\cdot)$, (3.38) implies that

$$\widehat{D}_t^q(\cdot) \to 0, \ \text{as } t \to \infty, \ in \ law. \tag{3.39}$$

Since the limit distribution is concentrated at zero, one may strengthen (3.39), to say that

$$\widehat{D}_t^q(\cdot) \to 0, \ \text{as } t \to \infty, \ in \ probability, \tag{3.40}$$

which implies that

$$\sup_{t \leq s \leq t+1} \|\widehat{D}_s^q\| \xrightarrow{P} 0, \ \text{as } t \to \infty. \tag{3.41}$$

A Localization Argument

To establish (3.41) for $\{\widehat{D}_s, t \leq s \leq t + 1\}_{t \geq 0}$, consider the following (standard) localization argument (see, e.g., Revus and Yor [38, Proof of Theorem IV-1.8]): Fix small $\varepsilon, \ \delta > 0$ and let $q < \infty$ be such that $P(\tau_q < \infty) \leq \delta$. Then,

$$P(\sup_{t \leq s \leq t+1} \|\widehat{D}_s\| > \varepsilon) \leq \delta + P(\sup_{t \leq s \leq t+1} \|\widehat{D}_s^q\| > \varepsilon), \tag{3.42}$$

which, since holding for all $\varepsilon, \delta > 0$, leads to

$$\sup_{t \leq s \leq t+1} \|\widehat{D}_s\| \xrightarrow{P} 0, \ \text{as } t \to \infty, \tag{3.43}$$

in turn implying the existence of a subsequence $\{r_n\}$ such that

$$\lim_{n \to \infty} \sup_{r_n \le s \le r_n + 1} \|\widehat{D}_s\| = 0, \ a.s. \tag{3.44}$$

A Contradiction

Assume that $P_t \nrightarrow 0$, namely that $\lambda_{max}\{P_\infty\} > 0$, w.p.p. (with positive probability). Let $\Gamma_t = \widehat{D}_t + P_t$. Then, due to (3.23) and (3.44),

$$\lim_{n \to \infty} \sup_{r_n \le s \le r_n + 1} \|\Gamma_s \Gamma_s^T - P_\infty^2\| =$$

$$= \lim_{n \to \infty} \sup_{r_n \le s \le r_n + 1} \left\| \Gamma_s \Gamma_s^T - U \begin{bmatrix} \Lambda^2 & \vdots & 0 \\ \cdots\cdots & & \cdots\cdots \\ 0 & \vdots & 0 \end{bmatrix} U^T \right\| = 0, \ w.p.p., \tag{3.45}$$

with $\Lambda = diag\{\lambda_1, \lambda_2 \cdots \lambda_r\}$, $r\varepsilon[1, 2, \cdots n^2]$ and $U^T = U^{-1}$ a unitary matrix.

Given (3.45), then, for any sufficiently small $\varepsilon > 0$, there exists an $L = L(\varepsilon, \omega) < \infty$ (a.s.) such that

$$\Gamma_s \Gamma_s^T \ge P_\infty^2 - \begin{bmatrix} \varepsilon I_r & \vdots & 0 \\ \cdots\cdots & & \cdots\cdots \\ 0 & \cdots & 0 \end{bmatrix} \ge 0, \ \forall \ s\varepsilon[r_n, r_n + 1], \ n \ge L, \ w.p.p. \tag{3.46}$$

Specifically, take $0 < \varepsilon < \min_{1 \le j \le r} \lambda_j^2/2$. Then, with the aforementioned subsequence $\{t_n\}$, (3.36) and (3.46) give

$$0 = \liminf_{n \to \infty} \frac{1}{t_n} \mathrm{Tr}\{\int_0^{t_n} \Gamma_s^T X_s^T X_s \Gamma_s ds\} = \liminf_{n \to \infty} \frac{1}{t_n} \mathrm{Tr}\{\int_0^{t_n} X_s \Gamma_s \Gamma_s^T X_s ds\} \ge$$

$$\ge \liminf_{n \to \infty} \frac{1}{n} \mathrm{Tr} \left\{ \sum_{i=L}^n \int_{r_i}^{r_i+1} X_s (P_\infty^2 - \begin{bmatrix} \varepsilon I_r & \vdots & 0 \\ \cdots\cdots & & \cdots\cdots \\ 0 & \vdots & 0 \end{bmatrix}) X_s^T ds \right\}$$

$$= \liminf_{n \to \infty} \frac{1}{n} \mathrm{Tr} \left\{ \sum_{i=L}^n U^T \int_{r_i}^{r_i+1} X_s^T X_s ds U \begin{bmatrix} \Lambda^2 - \varepsilon I_r & \vdots & 0 \\ \cdots & & \cdots\cdots \\ 0 & & \vdots 0 \end{bmatrix} \right\} =$$

$$= \liminf_{n \to \infty} \frac{1}{n} \sum_{i=L}^n \sum_{j=1}^r \left[U^T \int_{r_i}^{r_i+1} X_s^T X_s ds U \right]_{(jj)} (\lambda_j^2 - \varepsilon) > 0, \ w.p.p.$$

$$\tag{3.47}$$

The contradiction between the last inequality and the first equality, due to the choice of ε and the PE property (3.21), completes the proof that, subject to (3.37), $P_\infty = 0$, a.s.

Proof of the Hypothesis (3.37) To verify (3.37) we resort to a localization argument similar to that employed above (following (3.37) up to (3.43)): With $\{m_t = \int_0^t D_s dw_s,\ t \geq 0\}$ the (local) martingale part of $\{\alpha_t^*,\ t \geq 0\}$, consider the square integrable martingale $\{m_t^k,\ t \geq 0\}$ defined as $m_t^k = m_{t \wedge s_k}$, with $s_k = \inf\{t \geq 0 : \| < m >_t \| > k\}$ making

$$\sup_{t \geq 0} \int_t^{t+1} d < m^k >_s = \sup_{t \geq 0} \int_t^{t+1} D_s^k D_s^{kT} ds \leq kI, a.s.,\ I\varepsilon\mathbb{R}^{n \times n}. \tag{3.48}$$

With $< m >_t = < \alpha^* >_t$, it follows from (3.33) and (3.48) that

$$\mathrm{E}[\int_t^{t+1} d < m^k >_s] = \mathrm{E}[\int_t^{t+1} D_s^k D_s^{kT} ds]$$

$$= \mathrm{E}[\int_t^{t+1} \mathrm{E}[D_s^k D_s^{kT} | \mathcal{F}_s^x] ds] \to 0 \text{ as } t \to \infty, \tag{3.49}$$

giving

$$\int_t^{t+1} \widehat{D}_s^k \widehat{D}_s^{kT} ds \leq \int_t^{t+1} \mathrm{E}[D_s^k D_s^{kT} | \mathcal{F}_s^x] ds \xrightarrow{P} 0, \text{ as } t \to \infty, \tag{3.50}$$

where the inequality is a Jensen inequality.

Finally, fix $\varepsilon, \delta > 0$ and let $k < \infty$ be such that $P(s_k < \infty) \leq \delta$. Then,

$$P\left(\| \int_t^{t+1} \widehat{D}_s \widehat{D}_s^T ds \| > \varepsilon\right) \leq \delta + P\left(\| \int_t^{t+1} \widehat{D}_s^k \widehat{D}_s^{kT} ds \| > \varepsilon\right). \tag{3.51}$$

Since this holds for all $\varepsilon, \delta > 0$, then due to (3.50), conjecture (3.37) is established, which completes the proof of the theorem. $\square$

To conclude this section, the next theorem shows that AML estimates possess the same limit set as RLS estimates do: The main result of this section, establishing properties (i), (ii), and (iii) above for AML estimates, is the following.

Theorem 3.4 ([8, Theorem 3.4]) *Consider the system (2.1) with $C = I$ and $[A, B] = [A^*, B^*]$ a stabilizable (deterministic) pair. Suppose that $\Phi_t > 0$ a.s. for all $t \geq t_0$, for some $t_0 < \infty$, and let $\{\theta_t, t \geq t_0\}\varepsilon\mathcal{AML}$ in the sense of (3.3) with $\theta_t\varepsilon\mathcal{S}$ for all $t\varepsilon[t_0, \infty]$, a.s., where, in addition, $\{\theta_t\}$ is incorporated in an LQ feedback control, characterized by a feedback matrix $-K(\theta_t)$. Then $\theta_\infty = \lim_{t \to \infty} \theta_t$ exists and is finite, w.p.1, where*

$$\theta_\infty \varepsilon \mathcal{I} \stackrel{\Delta}{=} \{\theta = col(A, B) : A - BK(\theta) = A^* - B^* K(\theta)\}, \quad a.s. \tag{3.52}$$

This holds for all stabilizable pairs $[A^*, B^*]$ outside a Lebesgue null set $\mathcal{N}$ in $\mathbb{R}^{n \times n} \times \mathbb{R}^{n \times m}$.

Proof

An Auxiliary RLS Estimate

Assume throughout that $\theta^* = col(A^*, B^*)$ lies outside the Lebesgue null set $\mathcal{N}$ cited in Lemma 3.1. Let Θ_t be the matrix counterpart of the AML estimate θ_t (i.e., $\theta_t = col\Theta_t$), and consider a collection of RLS (matrix) estimates $\{\Theta_t^\varepsilon, t \geq 0\}_{0 < \varepsilon < 1}$, sharing a common initial estimate Θ_0, but different initial covariances. More precisely, let Θ_t^ε be an RLS (matrix) estimate with its resulting estimation error $\widetilde{\Theta}_t^\varepsilon = \Theta_t^\varepsilon - \Theta^*$ having the following closed form

$$\widetilde{\Theta}_t^\varepsilon = [Q_t + \varepsilon I]^{-1} [M_t + \varepsilon \widetilde{\Theta}_0], \tag{3.53}$$

where, with $\phi_t = [x_t^T, u_t^T]^T$, $M_t \stackrel{\Delta}{=} \int_0^t \phi_s dw_s^T$ and $Q_t \stackrel{\Delta}{=} \int_0^t \phi_s \phi_s^T ds$ are respectively a martingale matrix and its corresponding increasing process, where $col\Theta_t^\varepsilon$ is generated by algorithm (3.8, 3.9) with initial conditions $\theta_0 = col\Theta_0$ and $R_0 = \frac{1}{\varepsilon} I$.

Next, as in (3.11, 3.12), the $\mathcal{AML}$ estimation error $\widetilde{\Theta}_t$ may be written as

$$\widetilde{\Theta}_t = Q_t^{-1} [M_t - \nabla \Lambda_t(\Theta_t)] = \widetilde{\Theta}_t^\varepsilon + \varepsilon Q_t^{-1} [\widetilde{\Theta}_t^\varepsilon - \widetilde{\Theta}_0] - Q_t^{-1} \nabla \Lambda_t(\Theta_t), \tag{3.54}$$

where $\nabla \Lambda_t(\Theta) = M_t - Q_t[\Theta - \Theta^*]$ is the (matrix) log-likelihood gradient.

By the fact that $\theta_t \varepsilon \mathcal{F}_t^x$, $t \geq 0$, we see that $\widetilde{\alpha}_t^\varepsilon = col\{(\widetilde{\Theta}_t^\varepsilon)^T [I, -K^T(\theta_t)]^T\}$ is the RLS estimation error vector of the closed -loop dynamics. Then, taking the transpose of (3.54) and multiplying both sides from the right by $[I, -K^T(\theta_t)]^T$, the vector (or column) version of the resulting equation becomes

$$\widetilde{\alpha}_t \stackrel{\Delta}{=} col\{\widetilde{\Theta}_t^T [I, -K^T(\theta_t)]^T\} = \widetilde{\alpha}_t^\varepsilon + \varepsilon col\{[\widetilde{\Theta}_t^\varepsilon - \widetilde{\Theta}_0]^T Q_t^{-1} [I, -K^T(\theta_t)]^T\}$$
$$+ col\{\nabla \Lambda(\Theta_t)^T Q_t^{-1} [I, -K^T(\theta_t)]^T\}.$$

Convergence of the AML Estimate

Recall that by Lemma 3.1, $\theta_\infty = \lim_{t \to \infty} \theta_t$ exists (a.s.) and is finite. Here by hypothesis we also have $\theta_\infty \varepsilon \mathcal{S}$, a.s. Therefore, with the continuity of K,

$$K(\theta_t) \to K(\theta_\infty) \text{ a.s. as } t \to \infty. \tag{3.55}$$

Hence, by Corollary 3.3 and the definition of $\widetilde{\alpha}_t^\varepsilon$,

$$\widetilde{\alpha}_t^\varepsilon \to 0, \text{ a.s., as } t \to \infty, \forall \varepsilon, 0 < \varepsilon < 1. \tag{3.56}$$

Consequently, since in addition $\nabla \Lambda_t(\Theta_t) \to 0$, *a.s.* (by definition (3.6)), one has

$$\lim_{t \to \infty} \tilde{\alpha}_t = \lim_{t \to \infty} \tilde{\alpha}_t(\tilde{\theta}_t, \theta^*) = \tilde{\alpha}_\infty = \varepsilon col\{[\tilde{\Theta}_\infty^\varepsilon - \tilde{\Theta}_0]^T Q_\infty^{-1}[I, -K^T(\theta_\infty)]^T\}, \text{ a.s.}$$

$$(3.57)$$

(as Q_t^{-1} exists for all t sufficiently large and is nonincreasing). Note that the third term on the RHS of (3.54) goes to zero by the log-likelihood gradient property, which is the very definition of the class of $\mathcal{AML}$ estimates. This holds for all $\varepsilon, 0 < \varepsilon < 1$.

AML Closed-Loop Consistency

Our next step is to show that in fact $\tilde{\alpha}_\infty = 0$, by taking $\varepsilon \to 0$ (along a subsequence, $\varepsilon_n \to 0$, *as* $n \to \infty$). Note that the RHS term of (3.57), multiplied by ε, has only $\tilde{\Theta}_\infty^\varepsilon$ as an ε dependent term. Therefore it suffices to show that $\tilde{\Theta}_\infty^\varepsilon$ is uniformly bounded in ε. Toward this end, recall that

$$\tilde{\Theta}_t = Q_t^{-1}M_t - Q_t^{-1}\nabla\Lambda_t(\Theta_t).$$

Using this and the identity

$$[Q + \varepsilon I]^{-1} = Q^{-1} - \varepsilon[Q + \varepsilon I]^{-1}Q^{-1},$$

one may rewrite $\tilde{\Theta}_t^\varepsilon$ in (3.53) as

$$\begin{aligned}
\tilde{\Theta}_t^\varepsilon &= [Q_t^{-1} - \varepsilon[Q_t + \varepsilon I]^{-1}Q_t^{-1}][M_t + \varepsilon\tilde{\Theta}_0] \\
&= \tilde{\Theta}_t + Q_t^{-1}\nabla\Lambda_t(\Theta_t) + \varepsilon Q_t^{-1}\tilde{\Theta}_0 - \varepsilon[Q_t \\
&\quad + \varepsilon I]^{-1}[\tilde{\Theta}_t + Q_t^{-1}\tilde{\Theta}_0 + Q_t^{-1}\nabla\Lambda_t(\Theta_t)].
\end{aligned}$$

Take $t \to \infty$ and recall that $\tilde{\Theta}_t \to \tilde{\Theta}_\infty$ a.s., $Q_t^{-1}\nabla\Lambda_t(\Theta_t) \to 0$ a.s. and that $[Q_t + \varepsilon I]^{-1}$ is nonincreasing and uniformly bounded in ε and t; hence

$$\liminf_{t \to \infty} \sup_{0 < \varepsilon < 1} \|\Theta_t^\varepsilon\| < \infty \text{ a.s.}$$

Taking $\varepsilon \to 0$ (along a subsequence) for the RHS term in (3.57) leads to

$$\tilde{\alpha}_\infty = \tilde{\alpha}_\infty(\Theta^*) = 0, \text{ a.s.,}$$

for all stabilizable pairs $\Theta^* = [A^*, B^*]^T$, outside the Lebesgue null set $\mathcal{N}$ in $\mathbb{R}^{n \times n} \times \mathbb{R}^{n \times m}$. Then, by the definition of $\tilde{\alpha}_t(\Theta^*)$, we have

$$\tilde{\alpha}_\infty(\Theta^*) = col\{A_\infty - A^* - [B_\infty - B^*]K(\theta_\infty)\} = 0,$$

which completes the proof.   □

Remark Note that $\tilde{\alpha}_\infty = 0$ implies that the RHS of (3.57) is identically zero, for all ε, $0 < \varepsilon < 1$. In other words that

$$Q_t^{-1}[I, -K^T(\theta_t)]^T \to 0 \quad \text{a.s. as } t \to \infty.$$

To illustrate this fact, consider the simple case where the dimensions $m = n = 1$, $\theta_t = \theta = \text{const.}$, for all $t \geq t_0 > 0$ and $Q_{t_0} = \delta I$, $\delta > 0$. Let $k = k(\theta)$ be the (scalar) constant gain and define $p_t = \int_0^t x_s^2 ds$. Then

$$Q_t^{-1}[1, -k]^T = \begin{bmatrix} p_t + \delta , & -kp_t \\ -kp_t , & k^2 p_t + \delta \end{bmatrix}^{-1} \begin{bmatrix} 1 \\ -k \end{bmatrix}$$

$$= \frac{1}{\delta + p_t(1 + k^2)} \begin{bmatrix} 1 \\ -k \end{bmatrix} \to 0 \text{ as } t \to \infty$$

since $p_t \to \infty$ (due to (3.21)).

3.3   An Alternative Definition of the Limit Set $\mathcal{I}$

The following provides an alternative, data-based interpretation to the limit set $\mathcal{I}$:

Theorem 3.5 *Under the conditions of Theorem 3.4, the limits θ_∞ of $\mathcal{AML}$ estimates $\{\theta_t, \ t \geq t_0\}$ are characterized by the property*

$$\theta_\infty \varepsilon \{\theta \varepsilon S : \theta - \theta^* \varepsilon Ker(\overline{\Phi}(\theta))\} = \mathcal{I}(\theta^*) = \mathcal{I}, \quad (3.58)$$

where

$$\overline{\Phi}(\theta_\infty) = \lim_{t \to \infty} \frac{1}{t} \Phi_t = BlockDiag^{(n)} \left\{ \begin{bmatrix} I \\ -K(\theta_\infty) \end{bmatrix} \overline{P}(\theta_\infty)[I, \ -K^T(\theta_\infty)] \right\}$$
$$(3.59)$$

and

$$\overline{P}(\theta_\infty) = \lim_{t \to \infty} \frac{1}{t} \int_0^t x_s x_s^T ds = \int_0^\infty [exp F(\theta_\infty)t] CC^T [exp F^T(\theta_\infty)t] dt > 0, \ a.s.,$$
$$(3.60)$$

with $BlockDiag^{(n)}\{G\}$ being a block-diagonal matrix whose n blocks are the matrices G and where $F(\theta) = A - BK(\theta)$.

Proof From the definitions of Φ_t, Ψ_t, one may explicitly write

$$\frac{1}{t}\Phi_t = \frac{1}{t}\int_0^t \Psi_s^T \Psi_s \, ds$$

$$= BlockDiag^{(n)}\{G_t\}, \quad G_t = \frac{1}{t}\int_0^t \begin{bmatrix} I \\ -K(\theta_s) \end{bmatrix} x_s x_s^T [I, \ -K^T(\theta_s)] ds. \quad (3.61)$$

We recall from Lemma 3.1 that $\theta_t \rightarrow \theta_\infty$ (a.s.), and hence, due to continuity, $K(\theta_t) \rightarrow K(\theta_\infty)$, a.s. It therefore follows that

$$G_t \rightarrow G(\theta_\infty) = \begin{bmatrix} I \\ -K(\theta_\infty) \end{bmatrix} \lim_{t\to\infty} \frac{1}{t}\int_0^t x_s x_s^T \, ds [I, \ -K^T(\theta_\infty)], \ \text{as } t \rightarrow \infty, \ \text{a.s.}$$

$$(3.62)$$

With the limit $G(\theta_\infty)$, a.s. finite, positive semi-definite. This leads to

$$\overline{\Phi}(\theta_\infty) = BlockDiag^{(n)}\{G(\theta_\infty)\}, \quad (3.63)$$

where the existence of the limit on the RHS of (3.62) is substantiated in Chap. 6 (Theorem 6.2) with its explicit form

$$\lim_{t\to\infty} \frac{1}{t}\int_0^t x_s x_s^T \, ds = \overline{P}(\theta_\infty) = \int_0^\infty [exp F(\theta_\infty)t] CC^T [exp F^T(\theta_\infty)t] dt > 0, \ a.s.$$

$$(3.64)$$

Consider now the log-likelihood gradient, given by (3.1); necessarily,

$$\frac{1}{t}\nabla L_t(\theta_t) = \frac{1}{t}\int_0^t \Psi_s^T dw_s - \frac{1}{t}\Phi_t(\theta_t - \theta^*) \rightarrow 0, \ a.s., \ as \ t \rightarrow \infty, \quad (3.65)$$

but by (3.61, 3.62), the fact that $\Psi_s \varepsilon \mathcal{F}_s^x$, $\forall s \geq 0$, and the martingale Law of Large Numbers, itself based on (3.64), it follows that

$$\frac{1}{t}\int_0^t \Psi_s^T dw_s \rightarrow 0, \ a.s., \ as \ t \rightarrow \infty, \quad (3.66)$$

which, together with (3.65), results in

$$\lim_{t\to\infty} \frac{1}{t}\Phi_t(\theta_t - \theta^*) = \overline{\Phi}(\theta_\infty)(\theta_\infty - \theta^*) = 0, \quad (3.67)$$

with $\overline{\Phi}(\theta_\infty)$ given by (3.62, 3.63). Equation (3.67) immediately results in

$$\theta_\infty - \theta^* \varepsilon Ker(\overline{\Phi}(\theta_\infty)) = Ker\left(BlockDiag^{(n)}\{[I, \ -K^T(\theta_\infty)]\}\right), \quad (3.68)$$

and it can be easily verified that with $\theta - \theta^* = col\{(A - A^*), (B - B^*)\}$, one has

$$BlockDiag^{(n)}\{[I, \ -K^T(\theta)]\}(\theta - \theta^*) = col\{A - A^* - (B - B^*)K(\theta)\} = 0 \Leftrightarrow \theta \varepsilon \mathcal{I}.$$
$$(3.69)$$

$\square$

3.4   A Class of AML Estimates

We end this section with a specified class of AML estimates making $\nabla L_t(\theta_t)$ to decay exponentially fast. This follows a similar feature of imposed exponential decay as given in Levanony, Shwartz, and Zeitouni [30].

Lemma 3.6 ([8, Lemma 3.5]) *Fix some $\alpha > 0$, $\theta_0 \varepsilon \mathcal{S}$ and let $\{\theta_t, \ t \geq t_0\}$ be an estimate generated by the SDE,*

$$d\theta_t = \Phi_t^{-1}\{\Psi_t^T dx_t - [\Psi_t^T \Psi_t \theta_t - \alpha \nabla L_t(\theta_t)]dt\}, \ t \geq t_0, \ \theta_{t_0} = \theta_0, \qquad (3.70)$$

with $t_0 > 0$ being such that $\Phi_{t_0} > 0$. Then $\{\theta_t, \ t \geq t_0\}$ is an AML estimate with the property

$$\nabla L_t(\theta_t) = \nabla L_{t_0}(\theta_0) \exp^{-\alpha(t-t_0)}. \qquad (3.71)$$

Proof Given (3.70), we use Ito's law and the linear function $\nabla L_t(\theta)$ whose explicit form (3.1) is given by

$$\nabla L_t(\theta) = \int_0^t \Psi_s^T dx_s - \Phi_t \theta, \qquad (3.72)$$

where $\Phi_t = \int_0^t \Psi_s^T \Psi_s ds$, to obtain,

$$d\nabla L_t(\theta_t) = \Psi_t^T dx_t - \Psi_t^T \Psi_t \theta_t dt - \Phi_t d\theta_t = -\alpha \nabla L_t(\theta_t)dt, \qquad (3.73)$$

which, together with initial conditions at t_0, gives (3.71). $\square$

Chapter 4
Geometric Results

Abstract A geometric study of parameter estimates' limit sets of indistinguishable closed-loop dynamics is presented. It is shown that optimal LQ performance is conditional upon consistent parameter estimation. This is performed by showing that the parameterized, long-run quadratic cost has a unique minimum at the true parameter generating the observations. The projection is then defined of the control cost gradient onto the tangent space to the limit set of parameters associated with indistinguishable closed-loop dynamics. That projection is shown to have a unique stationary point at the true parameter. A scalar example is provided.

Keywords Geometric study · Unique minimum · Unique stationary point · Asymptotically stable · Projected gradient · Scalar example

In this section we examine the geometric characteristics of the limit set $\mathcal{I}$, the set of systems with indistinguishable closed-loop dynamics. As is apparent from Theorem 3.4, standard CE LQ schemes may only lead to parameter estimate convergence into $\mathcal{I}$, producing suboptimal performance. To deal with this situation, the geometric study in this section provides information which facilitates the construction of an adaptive control scheme which will result in the desired optimal performance.

4.1 A Conflict Between Identification and Control

Let

$$\mathcal{C} = \{\theta \in \mathcal{S}; K(\theta) = K(\theta^*)\}.$$

Since $\theta_t \to \mathcal{I}$, *as* $t \to \infty$, one would achieve the optimal performance (2.8) if, further, $\mathcal{I} \subset \mathcal{C}$. However, Polderman [34, 35] showed that:

© The Author(s), under exclusive license to Springer Nature Switzerland AG 2024

D. Levanony, P. E. Caines, *Stochastic Lagrangian Adaptation*, SpringerBriefs in Mathematics, https://doi.org/10.1007/978-3-031-73758-9_4

(i) $\mathcal{I} \cap \mathcal{C} = \{\theta^*\}$

(ii) $V(\theta) \geq V(\theta^*) \qquad \forall \theta \in \mathcal{I}$

where we recall that V is the solution of ARE (2.6). We show below that without consistent identification (i.e., $\theta_t \rightarrow \theta^*$) only suboptimal performance may be achieved. Due to (i), by using standard $\mathcal{AML}$ estimates, one may encounter

$$\theta_t \rightarrow \theta_\infty \neq \theta^* \tag{4.1}$$

thus obtaining $J(\theta_\infty) \geq J(\theta^*)$ (where, as defined in (2.7), $J(\theta)$ is the optimal achievable performance for a system parameterized by θ). We now examine the first-order derivatives of V and J on $\mathcal{I}$. We show that, in addition of being the unique minimum of J over $\mathcal{I}$, θ^* is also a unique stationary value point of the gradient of J with respect to B (Lemma 4.1). Such a result is important as a key tool in establishing the consistency of various gradient and Newton type algorithms including the one used in this work.

Note that for any $\theta \in \mathcal{I}$, the calculation of $V(\theta)$ can be made either by the ARE (2.6) or by

$$[A^* - B^*K(\theta)]^T V + V[A^* - B^*K(\theta)] + K^T(\theta)K(\theta) + I = 0, \tag{4.2}$$

where $K(\theta) = B(\theta)^T V(\theta) = B^T V(\theta)$. Further note that by (4.2) $V(\theta) = V(A, B)$ is actually a function of B only (where Definition (3.52) determines the corresponding A for all $\theta \in \mathcal{I}$).

4.2   The Unique Stationary Point of J over $\mathcal{I}$

Let $dJ(\theta)/dB$ be an $n \times m$ matrix whose typical (i, j) entry is $dJ(\theta)/dB_{ij}$. As just noted, for any $\theta \in \mathcal{I}$, $J(\theta) = \mathrm{Tr}V(\theta)$ is a function of B only, with $A = A(B)$, $\forall \theta = col\{(A, B)\} \in \mathcal{I}$. One therefore has

$$\frac{dJ(\theta)}{dB_{ij}} = \frac{\partial J(\theta)}{\partial B_{ij}} + \sum_{p,q=1}^{n} \frac{\partial A_{pq}}{\partial B_{ij}} \frac{\partial J(\theta)}{\partial A_{pq}}. \tag{4.3}$$

Lemma 4.1 ([8, Lemma 4.1])   *It holds that θ^* is the only stationary value point of dJ/dB over $\mathcal{I}$, that is,*

$$\frac{dJ(\theta)}{dB} = 0 \Leftrightarrow \theta = \theta^*, \quad \theta \in \mathcal{I}. \tag{4.4}$$

Moreover, θ^* is the unique minimizer of J over $\mathcal{I}$.

Proof   Recall that for any $\theta \in \mathcal{I}, V(\theta) = V(A, B)$ is the solution of the ARE equation (4.2). Fix $\theta \in \mathcal{I}$, define the $n \times n$ matrices $V'(i, j) = dV/dB_{ij}$, and let

$B'(i, j)$ be an $n \times m$ matrix whose (i, j) entry equals 1 and all the other entries are identically zero (i.e., $B'(i, j) = dB/dB_{ij}$). Differentiating the ARE (4.2) with respect to B_{ij} and using the fact that on $\mathcal{I}$, $A - BK = A^* - B^* K$ result in

$$[A - B\widetilde{K}]^T V'(i, j) + V'(i, j)[A - B\widetilde{K}] + V\Delta(i, j)V = 0, \tag{4.5}$$

where $\widetilde{K} = \widetilde{K}(\theta) \stackrel{\Delta}{=} B^{*^T} V(\theta)$ and $\Delta(i, j) \stackrel{\Delta}{=} B'(i, j)(B - B^*)^T + (B - B^*)B'^T(i, j)$.

Note that only the i-th row and the i-th column of the (symmetric) matrix $\Delta(i, j)$ are not identically zero, where

$$\Delta_{ip}(i, j) = \Delta_{pi}(i, j) = \begin{cases} (B - B^*)_{pj} \; ; & p \neq i \\ 2(B - B^*)_{ij} \; ; & p = i \end{cases}. \tag{4.6}$$

Now, suppose that $V'(i, j) = 0$. Then from (4.5) one has

$$V\Delta(i, j)V = 0, \tag{4.7}$$

which, since $V > 0$, implies that

$$\Delta(i, j) = 0 \quad \text{and hence,} \; B_{pj} = B^*_{pj} \quad p = 1, 2 \cdots n.$$

Conversely, on $\mathcal{I}$, $B = B^* \Rightarrow A = A^* \Rightarrow \widetilde{A} = A - B\widetilde{K}(\theta) = A^* - B^* K(\theta^*)$, which is asymptotically stable, and hence, (4.5) yields $V' = 0$ (as $\Delta = 0$). Therefore,

$$\frac{dV}{dB_{ij}} = 0 \; i = 1, \cdots n; j = 1, \cdots m \Leftrightarrow B = B^*, \; \theta \in \mathcal{I}, \tag{4.8}$$

and, since $\theta \in \mathcal{I}$, one also has $A = A^*$, and thus (4.8) is rewritten in an obvious notation as

$$\frac{dV(\theta)}{dB} = 0 \Leftrightarrow \theta = \theta^*, \; \theta \in \mathcal{I}. \tag{4.9}$$

We now show that the same holds for $dJ(\theta)/dB = \text{Tr}V(\theta)/dB$ on $\mathcal{I}$: Consider $dJ(\theta)/dB$; let $\theta \in \mathcal{I}$ and suppose that $\widetilde{A} \stackrel{\Delta}{=} A - B\widetilde{K}$ is asymptotically stable. Then $V'(i, j)$, the solution of (4.5), may be written in the following explicit form:

$$V'(i, j) = \int_0^\infty \exp\{\widetilde{A}^T t\} V\Delta(i, j)V \exp\{\widetilde{A}t\}dt. \tag{4.10}$$

Let $E(t) \triangleq V \exp\{\widetilde{A}t\}$. Then,

$$\frac{dJ(\theta)}{dB_{ij}} = \frac{d}{dB_{ij}} \text{Tr} V(\theta) = \text{Tr} V'(i, j) = \sum_{p=1}^{n} V'_{pp}(i, j)$$

$$= \sum_{p=1}^{n} \left[\int_0^\infty E^T(t) \Delta(i, j) E(t) dt \right]_{pp} = \sum_{p,r,\ell=1}^{n} \int_0^\infty E_{\ell p}(t) E_{pr}^T(t) dt \, \Delta_{r\ell}(i, j)$$

$$= 2 \sum_{r=1}^{n} \sum_{p=1}^{n} \int_0^\infty E_{ip}(t) E_{pr}^T(t) dt \, \Delta_{ir}(i, j)$$

$$= 2 \sum_{r=1}^{n} \left[\int_0^\infty E^T(t) E(t) dt \right]_{ir} \Delta_{ir}(i, j)$$

$$= 2 \left\{ \sum_{r=1}^{n} \left[\int_0^\infty E(t) E^T(t) dt \right]_{ir} (B - B^*)_{rj} \right.$$

$$\left. + \left[\int_0^\infty E(t) E^T(t) dt \right]_{ii} (B - B^*)_{ij} \right\}$$

$$\triangleq \{ [Z + \text{diag} Z][B - B^*] \}_{ij}, \tag{4.11}$$

where the last equalities follow from the definition of $\Delta(i, j)$ and the definitions,

$$\text{diag} Z \triangleq \text{diag} \{ Z_{11}, Z_{22} \cdots Z_{nn} \}, \ Z \triangleq V \int_0^\infty \exp\{\widetilde{A}t\} \exp\{\widetilde{A}^T t\} dt \, V.$$

Let $\overline{Z} \triangleq Z + \text{diag} Z$. Then $V > 0$ implies $\infty > \overline{Z} > 0$ in case $\widetilde{A}$ is asymptotically stable, and hence

$$\forall \theta \in \mathcal{I} \text{ with } \widetilde{A} \text{ asymptotically stable}; \ \frac{d}{dB} J(\theta) = \overline{Z}(B - B^*) = 0 \Leftrightarrow \theta = \theta^*. \tag{4.12}$$

It remains to show that indeed $\widetilde{A} = A - B\widetilde{K}$ is asymptotically stable for all $\theta \in \mathcal{I}$.

Let (λ, e) denote an eigenvalue and its corresponding eigenvector of $\widetilde{A}$. Multiplying both sides of equation (4.5) by $\overline{e}^T$ and e (where $\overline{e}$ is the complex conjugate of e) results in

$$2Re\{\lambda\}\overline{e}^T V'(i, j)e + \overline{e}^T V \Delta(i, j) Ve = 0, \tag{4.13}$$

where e is chosen such that $\|e\| = 1$.

Now, suppose that there exists a $\theta \in \mathcal{I}$ such that $\widetilde{A}(\theta)$ is asymptotically unstable, with an eigenvalue λ,

$$\lambda = a + bj, \qquad Re(\lambda) = a \geq 0. \tag{4.14}$$

With e being the corresponding unit eigenvector, (4.13), with the aid of (4.10), is rewritten as

$$\bar{e}^T V \Delta(i, j) V e \left(2a \int_0^\infty e^{2at} dt + 1\right) = 0, \tag{4.15}$$

which can obviously hold only if $Re(\lambda) = a \leq 0$.

Now, the case $a = 0$ implies, through (4.15), that

$$\Delta(i, j) = 0, i = 1, \cdots n; \; j = 1, \cdots m, \tag{4.16}$$

which, by definition (4.6), results in $B = B^*$, and therefore (as $\theta \in \mathcal{I}$) one also has $A = A^*$, i.e., $\theta = \theta^*$. This leads to that $\widetilde{A} = \widetilde{A}(\theta^*) = A^* - B^* K(\theta^*)$, which is asymptotically stable, a fact which contradicts the assumption that $a = Re(\lambda(\widetilde{A})) = 0$.

Next, by Polderman's result [34] cited above, one has that $J(\theta) = \text{Tr}(V(\theta)) \geq J(\theta^*)$, for all $\theta \in \mathcal{I}$. This, together with the fact that θ^* has been shown to be the only stationary point of dJ/dB over $\mathcal{I}$, implies that indeed θ^* is the unique minimizer of J over $\mathcal{I}$. $\qquad\qquad\square$

Finally, let

$$\nabla J(\theta) = [(\frac{\partial J(\theta)}{\partial col A})^T, (\frac{\partial J(\theta)}{\partial col B})^T]^T. \tag{4.17}$$

We end this section with the presentation of the *projection* of $\nabla J(\theta)$ on the tangent space to $\mathcal{I}$ at a point $\theta \in \mathcal{I}$.

Recall that for any $\theta \in \mathcal{I}$, $A = A(B)$ is determined by (3.52). Let $D = D(\theta) = dcol(A)/dcol(B)$ be an $n^2 \times nm$ matrix whose entries are

$$D_{ij}(\theta) = \frac{\partial (col A)_i}{\partial (col B)_j}. \tag{4.18}$$

Then, $\nabla^{\mathcal{I}} J(\theta)$, denoting the projection of $\nabla J(\theta)$ on the tangent space to $\mathcal{I}$ at θ, is

$$\nabla^{\mathcal{I}} J(\theta) = \begin{bmatrix} D^T(\theta) \\ I \end{bmatrix} [D(\theta)D^T(\theta) + I]^{-1} \left[D(\theta)\frac{\partial J(\theta)}{\partial col A} + \frac{\partial J(\theta)}{\partial col B} \right], \tag{4.19}$$

where I is an $nm \times nm$ identity matrix. Note that

$$[D(\theta), I]\nabla^{\mathcal{I}} J(\theta) = D(\theta)\frac{\partial J(\theta)}{\partial col A} + \frac{\partial J(\theta)}{\partial col B} = col\frac{dJ(\theta)}{dB}. \tag{4.20}$$

Hence,

$$\nabla^{\mathcal{I}} J(\theta) = 0 \Leftrightarrow \frac{dJ(\theta)}{dB} = 0, \tag{4.21}$$

where, since $\theta \in \mathcal{I}$, (4.4) implies that $\theta = \theta^*$.

4.3   The Scalar Example

With $\theta = (a, b)^T$, the ARE (2.6), rewritten as

$$2(a - bk(\theta))v(\theta) + k^2(\theta) + 1 = 0, \ k(\theta) = bv(\theta), \tag{4.22}$$

leads through differentiation (with $v = J$), to that

$$\frac{\partial J}{\partial b} = -k\frac{\partial J}{\partial a}, \ \frac{\partial J}{\partial a} = \frac{-v}{a - bk} \neq 0, \ \forall \theta \in \mathcal{S}. \tag{4.23}$$

Differentiating (with respect to b) the $\mathcal{I}$—defining relation,

$$a - bk(\theta) = a^* - b^*k(\theta), \tag{4.24}$$

taking into account (4.23), results in the following implicit equation for $d = da/db$ ($\theta \in \mathcal{I}$):

$$d = (b - b^*)b(d - k)\frac{\partial v}{\partial a} + 2k - b^*v. \tag{4.25}$$

Equation (4.19) takes the form

$$\nabla^{\mathcal{I}} J(\theta) = \begin{bmatrix} d \\ 1 \end{bmatrix}\frac{1}{d^2 + 1}\frac{\partial J}{\partial a}(d - k), \ \theta \in \mathcal{S}. \tag{4.26}$$

From here, it is easily verifiable that, on $\mathcal{I}$,

$$\theta = \theta^* \Leftrightarrow d = k \Leftrightarrow \nabla^{\mathcal{I}} J(\theta) = 0. \tag{4.27}$$

Chapter 5
Lagrangian Adaptation

Abstract The stochastic Lagrangian adaptation scheme proposed in this work is derived: First, a time-varying constrained optimization problem is formulated. Given that the resulting necessary conditions may lead to a triviality, these conditions are replaced by associated conditions where the control cost gradient is replaced with its projection onto the tangent space of parameters giving rise to indistinguishable closed-loop dynamics. Next, the SDEs of consistent approximations to the solutions of the constrained optimization problem are derived and the existence of unique strong solutions established. Finally, these solutions are shown to make the generated AML estimates strongly consistent from which optimal long-run LQ performance follows. This chapter concludes with the computation of the Regret Rate of the proposed scheme.

Keywords Optimal adaptive LQ performance · Lagrangian adaptive control · Projected gradient · Derivative of the closed-loop dynamics matrix · Estimate and adjoint process flows · Forced exponential decay · Smoothed projected gradient · Unique strong solution · Bounded trajectories · Uniformly asymptotically stable equilibrium · Uniformly bounded away · Initialization time · Regret rate

The results quoted in the previous section show that for optimal adaptive LQ performance it is necessary to generate consistent parameter estimates. Sufficiency follows from:

Theorem 5.1 ([13, Theorem 2]) *Let* $\{\widehat{\theta}_t, \ t \geq 0\}$ *be a consistent estimate in the sense that*

$$\widehat{\theta}_t \rightarrow \theta^*, \ a.s. \ as \ t \rightarrow \infty,$$

where in addition, $\widehat{\theta}_t \in \mathcal{S}$*, for all* $t \geq T$*, for some sufficiently large, possibly random* $T < \infty$ *a.s. Then, with* $\{\widehat{\theta}_t, \ t \geq 0\}$ *incorporated within an adaptive feedback law* *of the form,*

$$u_t = -K(\widehat{\theta_t})x_t,$$

the resulting long-run LQ performance is optimal in the sense of (2.8), a.s.

This leads us to adopt a methodology related to the biased ML approach of Kumar [25] and Borkar [3] (see also Kumar and Becker [23], Kumar and Lin [24]). The techniques of the aforementioned authors invoke the minimization of a weighted sum of the log-likelihood function and the computed performance J of the controlled system.

In light of Theorem 3.4, the point of view adopted in this work is that, in the limiting case of an infinite observation sample, the control task is to minimize J over the parameterized system descriptions and parameterized controllers that satisfy the *constraint* given by the vanishing of the gradient of the log-likelihood function. Consequently, on the finite time interval, we seek a suitable approximation to this constrained optimization task.

In Kumar [26] (where finite parameter sets are considered) and Borkar [3], the minimization of the relevant weighted sum is assumed to occur instantaneously. Since our problem formulation involves uncountable parameter spaces, this is highly impractical in the present case. Consequently, the adaptive algorithm presented in this work involves stochastic differential equations whose solutions approximate the solution of the constrained optimization problems referred to above. The resulting parameter estimation algorithm part of the adaptive control scheme will be in the form of an implicitly biased ML-type algorithm.

5.1   Motivation and Sketch of the Conceptual Adaptive Scheme

Suppose that a positive scalar stochastic process $\{\delta_t; t \geq 0\}$ is given with δ_t monotonically decreasing to zero. Then we formulate the adaptive optimization problem as follows:

$$\text{minimize} \qquad \mathcal{J}(\theta, \theta),$$

$$\text{subject to} \qquad \begin{cases} \theta \in \mathcal{S} \\ \|\nabla L_t(\theta)\| \leq \delta_t, \end{cases} \tag{5.1}$$

where, with a slight change of notation, the first parameter, θ, appearing in the function $\mathcal{J}(\cdot, \cdot)$ denotes the system $\Xi(\cdot)$, while the second refers to the parametrization of the feedback control law $K(\theta)$. Hence $\mathcal{J}(\theta, \theta')$ denotes the LQ performance of the system $\Xi(\theta)$ subject to the control law $u_t = -K(\theta')x_t$, and we adopt the convention that $\nabla J(\theta, \theta)$ denotes the gradient of $\mathcal{J}(\theta, \theta)$ with respect to the θ parameter in both entries.

The constraint in (5.1) is introduced so as to make (by definition) the resulting solution of (5.1) an AML estimate. Such an estimate has been shown in Theorem 3.4 above to converge to a limit in the set $\mathcal{I}$, a set on which θ^* is the unique minimizer of the control cost $\mathcal{J}$ (Lemma 4.1). This fact leads to the formulation of (5.1), with the constraint placed so as to guarantee that indeed, asymptotically, minimization of $\mathcal{J}$ is restricted to $\mathcal{I}$, resulting, due to the second part of Lemma 4.1, in the limit $\theta_\infty = \theta^*$. This, together with Theorem 5.1, will then lead to optimal linear quadratic performance.

The starting point for the formulation of the Lagrangian Adaptive Control algorithm is the necessary solution of (5.1). This formulation culminates in the specification of the *recursive* algorithm equations. Suppose that $\theta_t^o \in \mathcal{S}$ is a solution to (5.1) (at time t), and then there exists an adjoint variable $\lambda_t^o \geq 0$ such that

$$\nabla \mathcal{J}(\theta_t^o, \theta_t^o) - \lambda_t^o \Phi_t \nabla L_t(\theta_t^o) = 0, \tag{5.2}$$

$$\|\nabla L_t(\theta_t^o)\| \leq \delta_t, \tag{5.3}$$

where we note that $\frac{1}{2} \frac{\partial}{\partial \theta} \|\nabla L_t(\theta)\|^2 = -\Phi_t \nabla L_t(\theta)$.

The scalar case shows $\nabla \mathcal{J}$ and $\Phi_t \nabla L_t$ to be asymptotically parallel for *any* $\theta \in \mathcal{S}$, and this may lead to an asymptotically trivial satisfaction of the necessary condition (5.2). Specifically, we observe that in the scalar case ($\theta = (a, b)^T$) (4.17) reduces to

$$\frac{\partial v}{\partial b} = -k \frac{\partial v}{\partial a}; \quad \frac{\partial v}{\partial a} = \frac{-v}{a - bk} > 0, \quad \forall \theta = (a, b)^T \in \mathcal{S}, \ k = k(\theta). \tag{5.4}$$

In this case $\mathcal{J} = v$ and so $\nabla \mathcal{J}(\theta, \theta)$ is parallel to the vector $(1, -k)^T$. However, by a simple calculation $\Phi_t \nabla L_t(\theta)$ converges to a vector parallel to $(1, -k)^T$, as is shown by

$$\lambda_t \Phi_t \nabla L_t(\theta) \to -c\overline{p}^2(1 + k^2(\theta)) \begin{bmatrix} 1 \\ -k \end{bmatrix} [a - a^* - k(b - b^*)], \ a.s., \ as \ t \to \infty, \tag{5.5}$$

where $\overline{p} = \lim_{t \to \infty} \frac{1}{t} \int_0^t x_s^2 ds$, $\lambda_t = c/t^2$, ($c > 0$).

By choosing the appropriate constant c, one may therefore obtain

$$\lim_{t \to \infty} (\nabla \mathcal{J}(\theta, \theta) - \lambda_t \Phi_t \nabla L_t(\theta)) = \lim_{t \to \infty} \left(\frac{\partial J}{\partial a} \begin{bmatrix} 1 \\ -k \end{bmatrix} - \lambda_t \Phi_t \nabla L_t(\theta) \right) = 0, \tag{5.6}$$

for *any* $\theta \in \mathcal{S} - \mathcal{I}$. This triviality essentially results from the fact that we have shown both terms in (5.6) to be parallel to the vector $(1, -k)^T$ throughout $\mathcal{S}$ (as $t \to \infty$). We note that the statement above excludes $\theta \in \mathcal{I}$, and in particular, θ^*: On $\mathcal{I}$, the $\mathcal{I}$-defining relation (4.24), together with (4.23), implies that, asymptotically, the left-hand side of (5.6) nowhere vanishes on $\mathcal{I}$, a fact that excludes $\mathcal{I}$, hence θ^*, from the possible limit set in this case. In light of Theorem 5.1 and Lemma 4.1, this fact implies that optimal LQ performance is impossible in this setting.

Furthermore, if the minimization of $\mathcal{J}(\theta^*, \theta)$ is approximated for θ^* by adopting a CE approach, that is, assuming $\theta^* = \theta_t$ (the current estimate), we obtain the triviality

$$\theta_t = \arg\min_{\theta \in \mathcal{S}} \mathcal{J}(\theta_t, \theta).$$

Henceforth, we use the original notation $J(\theta) \triangleq \mathcal{J}(\theta, \theta)$.

The property above leads to the reformulation of (5.2, 5.3), resulting in following set of equations:

$$\nabla^{\mathcal{I}} J(\theta) - \lambda \Phi_t \nabla L_t(\theta) = 0, \tag{5.7}$$

$$\|\nabla L_t(\theta)\| \le \delta_t, \tag{5.8}$$

where the definition (4.19) of $\nabla^{\mathcal{I}} J(\theta)$ is extended to apply to *all* θ's in $\mathcal{S}$ and $\nabla^{\mathcal{I}} J(\theta)$ denotes the projection of $\nabla J(\theta)$ onto the tangent space T'_θ to $\mathcal{I}$ at θ', where θ' is the nearest point to θ in $\mathcal{I}$. We note that since $\mathcal{I}$ is unknown, $\nabla^{\mathcal{I}}$ is not computable. Therefore a consistent approximation of $\nabla^{\mathcal{I}} J$ is introduced below.

Before doing that, we recall the scalar example outlined above, to write the analogue to the left-hand side of (5.7), so as to demonstrate that no triviality occurs with regard to that expression. This is contrary to the case of (5.2), shown to result in a triviality, as demonstrated in the scalar case, through (5.6). With the calculations in (4.26) and (5.5), the left-hand side of (5.7) in the scalar case takes the form (recall that $d = da/db$),

$$\lim_{t \to \infty} (\nabla^{\mathcal{I}} J(\theta) - \lambda_t \Phi_t \nabla L_t(\theta)) = \begin{bmatrix} d \\ 1 \end{bmatrix} \frac{1}{d^2+1} \frac{\partial J}{\partial a} (d - k)$$

$$+ c\overline{p}^2(1 + k^2(\theta)) \begin{bmatrix} 1 \\ -k \end{bmatrix} [a - a^* - k(b - b^*)]. \tag{5.9}$$

Hence, with (4.27) and (4.24), one concludes that, for $\theta \in \mathcal{I}$, (5.9) is identically zero if and only if $\theta = \theta^*$.

5.2 An Estimate of the B-Derivative of the Closed-Loop Matrix

We begin with an observation driven, consistent approximation result for the (full) derivative of the closed-loop dynamic matrix $F(\theta) \triangleq A - BK(\theta)$ with respect to B, for a t-limit parameter $\theta = \theta_\infty$ laying in $\mathcal{I}$; this result is proved in Chap. 6.

(Recall that for any $\theta \in \mathcal{I}$, $F(\theta) = F^*(\theta) \triangleq A^* - B^* K(\theta)$.) Specifically, let $P_t \triangleq \int_0^t x_s x_s^T ds$, $\Gamma_t \triangleq P_t^{-1} \int_0^t x_s dx_s^T$. With $\Gamma_t \to F^T(\theta_\infty)$, we show that the derivative of Γ_t (a derivative in a certain sense to be specified below) with respect to B_{ij} converges to $dF^T(\theta_\infty)/dB_{ij}$, where $F(\theta) = A - BK(\theta)$, $\theta = col(A, B)^T$.

Theorem 5.2 *Let $\{\theta_t\}$ be an $\mathcal{AML}$ estimate incorporated in a CE LQ adaptive control, and suppose that there exists $\theta_\infty = \theta_\infty(\omega) \in \mathcal{I}$ such that $\theta_t \to \theta_\infty$ a.s. Let $P_t \triangleq \int_0^t x_s x_s^T ds$, $\Gamma_t \triangleq P_t^{-1} \int_0^t x_s dx_s^T$. For a fixed pair (i, j), $i = 1, \cdots n; j = 1, \cdots m$, define the matrix $\Pi_t(i, j)$ and the vector $\zeta_t(i, j)$ by*

$$\Pi_t(i, j) = P_t^{-1}\Big\{ \int_0^t \zeta_s(i, j)dx_s^T + \int_0^t x_s d\zeta_s^T(i, j)$$

$$- \Big[\int_0^t \big(\zeta_s(i, j)x_s^T + x_s \zeta_s^T(i, j) \big)ds \Big]\Gamma_t \Big\}, \quad t \geq t_0, \tag{5.10}$$

$$\dot{\zeta}_t(i, j) = F(\theta_t)\zeta_t(i, j) + \Pi_t^T(i, j)x_t, \quad \zeta_0(i, j) = 0, \tag{5.11}$$

$$\text{with } \Pi_t(i, j) = \frac{\partial F^T(\theta_t)}{\partial B_{ij}}, \quad 0 \leq t < t_0,$$

where $F(\theta) = A - BK(\theta)$, $\theta = col(A, B)^T$. Then

$$\Pi_t(i, j) \to \frac{d}{dB_{ij}}F^T(\theta_\infty), \quad a.s. \text{ as } t \to \infty. \tag{5.12}$$

Proof See Chap. 6. □

Remark For $\theta \in \mathcal{I}$, the (full) differentiation of $F(\theta) = A - BK(\theta)$ with respect to B inherently involves the computation of dA/dB which is impossible to evaluate for all $t \geq t_0$, as the functional dependence of A on B is unknown throughout $[t_0, \infty)$. Moreover, the functional dependence of A on B only occurs at the asymptotic limit as $\theta_t \to \theta_\infty \in \mathcal{I}$, with that function determined by the definition of $\mathcal{I}$ (3.52) which depends upon θ^*. This motivates a data-based approximation given by (5.10, 5.11) which is shown by Theorem 5.2 to converge to the desired limit. That limit is the *full derivative* of $F(\theta)$ with respect to B evaluated at $\theta = \theta_\infty \in \mathcal{I}$, on which the (full) functional dependence of $F(\theta)$ on B is well defined by virtue of the definition of $F(\theta) = A - BK(\theta)$, the definition of $K(\theta)$ (2.5, 2.6) together with the definition of $\mathcal{I}$ (3.52) substantiating that $\theta = col(A, B) \in \mathcal{I} \Rightarrow A = A(B)$.

Next, note that for any $\theta \in \mathcal{I}$, $\lim_{t \to \infty} \Pi_t(i, j, \theta) = dF^T(\theta)/dB_{ij}$ satisfies

$$\lim_{t \to \infty} \Pi_t(i, j, \theta) = \Big(\frac{dA}{dB_{ij}} - \Big[\frac{dB}{dB_{ij}}B^T + B\frac{dB^T}{dB_{ij}} \Big]V(\theta) - BB^T\frac{dV(\theta)}{dB_{ij}} \Big)^T,$$

$$\tag{5.13}$$

where

$$\frac{dV_{pq}}{dB_{ij}} = \sum_{k=1}^{n} \sum_{\ell=1}^{n} \frac{\partial V_{pq}}{\partial A_{k\ell}} \frac{dA_{k\ell}}{dB_{ij}} + \frac{\partial V_{pq}}{\partial B_{ij}} \tag{5.14}$$

for $p, q = 1, 2 \cdots n$, $i = 1, 2 \cdots n$ and $j = 1, 2 \cdots m$.

By definition $D = dcol A/dcol B$, hence its $n^3 m$ entries correspond to the elements of the tensor $\{dA_{k\ell}/dB_{ij}\}$ (where $k, \ell, i = 1, \cdots n$ and $j = 1, 2 \cdots m$).

Therefore equations (5.13) and (5.14) (with $1 \le i \le n, 1 \le j \le m$) form a set of $n^3 m$ algebraic equations for the $n^3 m$ unknown entries of D. We claim the following:

Lemma 5.3 *Consider a fixed $\theta \in S$ and an arbitrary $n \times n$ matrix Π. Then, the set of n^2 equations*

$$\Pi = \left(X_{ij} - \left[\frac{dB}{dB_{ij}} B^T + B \frac{dB^T}{dB_{ij}} \right] V(\theta) - BB^T \frac{dV(\theta)}{dB_{ij}} \right)^T \tag{5.15}$$

$$\frac{dV_{pq}}{dB_{ij}} = \sum_{k=1}^{n} \sum_{\ell=1}^{n} \frac{\partial V_{pq}}{\partial A_{k\ell}} X_{ij}(k, \ell) + \frac{\partial V_{pq}}{\partial B_{ij}} \tag{5.16}$$

has a unique solution $X_{ij} = \{X_{ij}(k, \ell)\ k, \ell = 1, 2 \cdots n\}$, depending analytically on θ.

Remark Note that for a $\theta \in \mathcal{I}$, $\Pi = dF^T(\theta)/dB_{ij}$, and one has that $X_{ij}(k, \ell) = dA_{k\ell}/dB_{ij}$.

Proof Let

$$M = M(\theta) = A - BK(\theta) - (A^* - B^* K(\theta))$$

and define the set

$$\mathcal{I}_M = \{\theta' \in S : A' - B'K(\theta') - (A^* - B^* K(\theta')) = M\}.$$

Then, following Polderman [34, 35], $\mathcal{I}_M$ is an $n \times m$ dimensional C^ω-manifold, which implies that for any θ' in $\mathcal{I}_M, dA'/dB'_{ij}$ is uniquely defined and hence satisfies equations (5.15) and (5.16) (for a given $\Pi = \Pi(\theta')$).

In particular, for $\theta' = \theta$, dA/dB_{ij} is the unique solution of (5.15), (5.16) (for any given $\Pi = \Pi(\theta)$). This in turn implies that, with (5.15) and (5.16) symbolically written as

$$Hz = b,$$

where $z = dcol A/dB_{ij}$, $H = H(\theta)$, an $n^2 \times n^2$ *invertible* matrix, while the free vector b consists of (among other terms) the elements of Π. Hence, varying b by selecting Π would result in a unique solution $dA/dB_{ij}(\theta, \Pi)$. $\square$

Remark Note that for any $\theta \in \mathcal{I}$, $M(\theta) = 0$ and $\mathcal{I} = \mathcal{I}_M = \mathcal{I}_0$.

The Two Forms of dA/dB

For a $\theta = col(A, B) \in \mathcal{I}$, recall the definition of $D(\theta)$, the (full) derivative of $col(A)$ with respect to $col(B)$ whose entries are given by (4.18). Then, for a θ *not necessarily in* $\mathcal{I}$, the $n \times n$ matrices $\{X_{ij}, \ i = 1, 2 \cdots n, \ j = 1, 2 \cdots m\}$, solving (5.15, 5.16) (for given Π), are approximating $\{dA/dB_{ij}, \ i = 1, 2 \cdots n, \ j = 1, 2 \cdots m\}$ and may immediately be rearranged to form $\widehat{D}(\theta) \cong dcol(A)/dcol(B)$ the approximation of the derivatives D of (4.18). In the following, we therefore directly translate solutions X of (5.15, 5.16) for any given $\theta = col(A, B)$ to $\widehat{D}(\theta)$.

Corollary 5.4 *Given an estimate θ_t, let $X_{ij} = X_{ij}(\theta_t)$ be the unique solution of equations (5.15), (5.16) (with (i, j) ranging over $\{1, 2 \cdots n\}$ and $\{1, 2 \cdots m\}$, respectively), with $\Pi = \Pi_t(i, j, \theta_t)$ (see (5.10, 5.11)). Rearranging the entries of $\{X_{ij}(\theta_t), \ i = 1, 2 \cdots n, \ j = 1, 2 \cdots m\}$, to form $\widehat{D}_t(\theta_t)$ (described above), we may state the following.*

If there exists a $\theta_\infty \in \mathcal{I}$ such that

$$\theta_t \to \theta_\infty \text{ a.s. as } t \to \infty,$$

then it is the case that

$$\widehat{D}_t(\theta_t) \to D(\theta_\infty) \text{ a.s. as } t \to \infty \qquad (5.17)$$

and, with $\widehat{\nabla^{\mathcal{I}} J_t}(\theta)$ defined by (4.19) with $D(\theta)$ replaced by $\widehat{D}_t(\theta)$, it follows that

$$\widehat{\nabla^{\mathcal{I}} J_t}(\theta_t) \to \nabla^{\mathcal{I}} J(\theta_\infty) \text{ a.s. as } t \to \infty. \qquad (5.18)$$

Proof Follows directly from (5.12) and Lemma 5.3.   □

The last result leads to the following set of Lagrange-type equations whose solution (θ_t, λ_t) forms the foundation of the Lagrangian Adaptation procedure: First choose $\mathcal{F}_t^x$ measurable processes $0 < \delta_t \downarrow 0$ such that with $r_t \overset{\Delta}{=} \text{Tr}\Phi_t$, $r_t\delta_t \to 0$ as $t \to \infty$. Second, let (θ_t, λ_t) denote the solutions to the following set of equations:

$$\widehat{\nabla^{\mathcal{I}} J_t}(\theta_t) - \lambda_t \Phi_t \nabla L_t(\theta_t) = 0, \qquad (5.19)$$

$$\|\nabla L_t(\theta_t)\| \le \delta_t, \qquad (5.20)$$

where

$$\widehat{\nabla^{\mathcal{I}} J_t}(\theta) \overset{\Delta}{=} \begin{bmatrix} \widehat{D}_t^T(\theta) \\ I \end{bmatrix} \left[\widehat{D}_t(\theta)\widehat{D}_t^T(\theta) + I \right]^{-1} \left[\widehat{D}_t(\theta) \frac{\partial J(\theta)}{\partial col A} + \frac{\partial J(\theta)}{\partial col B} \right] \qquad (5.21)$$

and the entries of $\widehat{D}_t(\theta)$ are given by $\{X_{ij}(k, \ell), \quad i = 1, 2, \cdot, \cdots, n, \quad j = 1, 2, \cdot, \cdots, m, \; k, \ell = 1, 2, \cdot, \cdots, n\}$, which are the solutions of these quantities

$$\Pi_t(i, j) = \left(X_{ij} - \left[\frac{dB}{dB_{ij}}B^T + B\frac{dB^T}{dB_{ij}}\right]V(\theta) - BB^T\frac{dV(\theta)}{dB_{ij}}\right)^T \quad (5.22)$$

$$\frac{dV_{pq}(\theta)}{dB_{ij}} = \sum_{k=1}^{n}\sum_{\ell=1}^{n}\frac{\partial V_{pq}}{\partial A_{k\ell}}X_{ij}(k, \ell) + \frac{\partial V_{pq}(\theta)}{\partial B_{ij}}, \quad (5.23)$$

where $i = 1, 2 \cdots n$, $j = 1, 2 \cdots m$, $p, q = 1, 2 \cdots n$. As a direct derivation from (5.10), (5.11), Π_t evolves according to the SDE:

$$d\Pi_t(i, j) = P_t^{-1}\{\zeta_t(i, j)dx_t^T + x_t\zeta_t^T(i, j)F^T(\theta_t)dt$$

$$+ \int_0^t (\zeta_t(i, j)x_s^T + x_s\zeta_t^T(i, j))ds\,P_t^{-1}x_t[x_t^T\Gamma_t dt - dx_t^T]\}$$

$$- [\zeta_t(i, j)x_t^T + x_t\zeta_t^T(i, j)]\Gamma_t dt, \; t \geq t_0, \; \Pi_{t_0}(i, j), \quad (5.24)$$

$$\dot{\zeta}_t(i, j) = F(\theta_t)\zeta_t(i, j) + \Pi_t^T(i, j)x_t, \; \zeta_0(i, j) = 0, \; t \geq 0, \quad (5.25)$$

$$\text{with } \Pi_t(i, j) = \frac{\partial F^T(\theta_t)}{\partial B_{ij}}, \; 0 \leq t < t_0,$$

where the initial condition $\Pi_{t_0}(i, j)$, set at $t_0 > 0$, for which $P_{t_0} = \int_0^{t_0} x_s x_s^T ds \geq \varepsilon I$, for some preset $\varepsilon > 0$ (I being an $n \times n$ identity matrix), with $\Pi_{t_0}(i, j)$ given by (5.10) with $t = t_0$.

Obviously, a real-time application requires a *recursive* algorithm. Therefore, equations (5.19), (5.20) are to be used as a basis for the derivation of the SDEs (stochastic differential equations) for the evolution of their solutions (θ_t, λ_t) over time. As an asymptotic solution will evidently suffice, we shall not require (θ_t, λ_t) to satisfy (5.19) at any given finite time, but instead, to drive the LHS of (5.19) to zero as $t \to \infty$ (with a similar procedure applied for (5.20)). The resulting SDEs, termed as the Lagrangian equations, are derived below.

Fix some $\beta > 0$. Then, in order to avoid the need to compute derivatives of $\widehat{\nabla^\mathcal{I}}J_t(\theta)$, the proposed algorithm will use a *smoothed* version of this term of the form,

$$\gamma_t = \beta \int_0^t e^{-\beta(t-s)}\widehat{\nabla^\mathcal{I}}J_s(\theta_s)ds, \quad (5.26)$$

where here, the existence of process $\{\theta_t\}$ is assumed. The process $\{\gamma_t\}$ has the following asymptotic property:

Proposition 5.5 *If $\theta_t \to \theta_\infty \in \mathcal{I}$, a.s., as $t \to \infty$, then*

$$\gamma_t \to \nabla^\mathcal{I}J(\theta_\infty), \; \text{a.s., as } t \to \infty. \quad (5.27)$$

Proof The result follows immediately from Corollary 5.4.   □

5.3   Derivation of the Estimate and the Adjoint Processes Flows

The basic scheme is aimed to recursively generate approximate, asymptotically consistent solutions $\{\theta_t, \lambda_t\}$ to the (time varying) constrained optimization problem (5.1): Define

$$h_t(\theta, \lambda) = \gamma_t - \lambda \Phi_t \nabla L_t(\theta), \tag{5.28}$$

and the slack variable process

$$q_t(\theta) = \frac{1}{2}(\|\nabla L_t(\theta)\|^2 - \delta_t^2). \tag{5.29}$$

Fix $\alpha > \beta > 0$ and assume that there exist continuous semimartingales $(\theta, \lambda)_t$ such that

$$dh_t(\theta_t, \lambda_t) = -\beta h_t(\theta_t, \lambda_t)dt \tag{5.30}$$

$$dq_t(\theta_t) = -\alpha q_t(\theta_t).dt \tag{5.31}$$

Then, a *formal* application of a generalized Îto rule (Kunita [27, Theorem 3.3.1]), for $h_t(\cdot)$ and $q_t(\cdot)$, enables to obtain SDEs for (θ_t, λ_t), via an explicit presentation of the LHS of (5.30, 5.31). The issue of existence and uniqueness of solutions to the resulting SDEs is addressed in Theorem 5.6. (See a similar application in [30].)

Remark We note that the forced exponential decay of (h_t, q_t) implies that these quantities are only implicit functions of (θ_t, λ_t). Still, the resulting specified trajectories $\{\theta_t, \lambda_t, t \geq t_0\}$ derived below are such that indeed guarantee the aforementioned exponential decay.

In order to simplify the derivation to follow, we omit the explicit dependence upon $(\theta, \lambda)_t$ in most of the functions involved. Furthermore, we denote by

$$d\tilde{v}_t = dx_t - \Psi_t \theta_t dt \tag{5.32}$$

the increment process which we note is not a Brownian motion, unless $\theta = \theta^*$, and which hence may be called a pseudo-innovation. Further define

$$H_t = \frac{\partial h_t(\theta, \lambda)}{\partial \theta} = \lambda \Phi_t^2, \quad \dot{\Phi}_t = \Psi_t^T \Psi_t, \tag{5.33}$$

the pseudo-Hessian, and the derivative of the matrix process Φ_t (3.2), respectively.

We begin with the SDE for h:

$$dh_t = -\lambda_t \Phi_t \Psi_t^T d\tilde{v}_t - \lambda_t \dot{\Phi}_t \nabla L_t(\theta_t)dt + H_t d\theta_t + \Phi_t^2 d < \theta, \lambda >_t$$
$$-\Phi_t \Psi_t^T d < x, \lambda >_t -\Phi_t \nabla L_t(\theta_t)d\lambda_t + d\gamma_t = -\beta h_t dt. \qquad (5.34)$$

Now, from Îto's differentiation rule,

$$d\nabla L_t(\theta_t) = \Psi_t^T d\tilde{v}_t - \Phi_t d\theta_t, \qquad (5.35)$$

which leads to

$$dq_t = \nabla L_t^T(\theta_t)d\nabla L_t(\theta_t) + \frac{1}{2}\text{Tr}\{d < \nabla L(\theta) >_t\} - \delta_t d\delta_t$$

$$= \nabla L_t^T(\theta_t)\Psi_t^T d\tilde{v}_t - \nabla L_t^T(\theta_t)\Phi_t d\theta_t + \frac{1}{2}\text{Tr}\{\Phi_t^2 d < \theta >_t$$

$$+\Psi_t^T d < x >_t \Psi_t - 2\Psi_t \Phi_t d < \theta, x >_t\} - \delta_t d\delta_t = -\alpha q_t dt. \quad (5.36)$$

Assuming $H_t > 0$ and rearranging (5.34) result in

$$d\theta_t = H_t^{-1}\left\{\lambda_t \Phi_t \Psi_t^T d\tilde{v}_t + \lambda_t \dot{\Phi}_t \nabla L_t(\theta_t)dt - d\gamma_t - \beta h_t dt\right.$$

$$\left.-\Phi_t^2 d < \theta, \lambda >_t +\Phi_t \Psi_t^T d < x, \lambda >_t +\Phi_t \nabla L_t(\theta_t)d\lambda_t\right\}. \quad (5.37)$$

We define for later use

$$\mu_t(\theta, \lambda) = \lambda/\|\nabla L_t(\theta)\|^2. \qquad (5.38)$$

Introducing (5.37) into (5.36) and rearranging yield

$$d\lambda_t = \mu_t(\alpha q_t dt - \delta_t d\delta_t)$$

$$+\frac{1}{2}\mu_t \text{Tr}\{\Phi_t^2 d < \theta >_t +\Psi_t^T d < x >_t \Psi_t - \Psi_t \Phi_t d < \theta, x >_t\}$$

$$+\mu_t \nabla L_t^T(\theta_t)\Phi_t H_t^{-1}\left\{\beta h_t dt - \lambda_t \dot{\Phi}_t \nabla L_t(\theta_t)dt + \Phi_t^2 d < \theta, \lambda >_t\right.$$

$$\left.-\Phi_t \Psi_t^T d < x, \lambda >_t +d\gamma_t\right\}. \qquad (5.39)$$

Introducing (5.39) into (5.37), the diffusion-like term of the SDE for θ_t becomes $\Phi_t^{-1}\Psi_t^T d\tilde{v}_t$, which, together with the corresponding term in (5.39), enables one to compute the various quadratic and cross variations appearing in both (5.37) and (5.39), in terms of the quadratic variation of the observed process $\{x_t\}$. Let $\ell_t(\theta) = \Phi_t \nabla L_t(\theta)$. Then, these calculations, introduced into (5.37) and (5.39), result in the SDEs for $(\theta, \lambda)_t$:

$$d\lambda_t = \mu_t\{\alpha q_t dt - \delta_t d\delta_t\} + \mu_t \ell_t^T(\theta_t) H_t^{-1}\left\{d\gamma_t + \beta h_t dt - \lambda_t \dot{\Phi}_t \nabla L_t(\theta_t) dt\right\} \tag{5.40}$$

and

$$d\theta_t = \Phi_t^{-1}\Psi_t^T\{dx_t - \Psi_t \theta_t dt\} + \mu_t H_t^{-1}\ell_t(\theta_t)\{\alpha q_t dt - \delta_t d\delta_t\}$$
$$+ H_t^{-1}[I - \mu_t \ell_t(\theta_t)\ell_t^T(\theta_t) H_t^{-1}]\left\{\lambda_t \dot{\Phi}_t \nabla L_t(\theta_t) dt - d\gamma_t - \beta h_t dt\right\}, \tag{5.41}$$

where $\dot{\Phi}_t = \Psi_t^T \Psi_t$, with the identities $d < x >_t = I dt$, and $d < \theta >_t = \Phi_t^{-1}\Psi_t^T d < x >_t \Psi_t \Phi_t^{-1}$ being used.

Remark Note that if one takes only the first term on the RHS of (5.41), one obtains an (standard) ML algorithm. Hence, we say that (5.41) generates *biased* ML estimates.

Set $\delta_t = \exp\{-\alpha t/2\}$. With the unit vector $e_t = \nabla L_t(\theta_t)/\|\nabla L_t(\theta_t)\|$, together with the definitions of the terms $\mu_t = \lambda_t/\|\nabla L_t(\theta_t)\|^2$ and $\ell_t = \Phi_t \nabla L_t(\theta_t)$, the λ_t equation (5.40) may be further simplified to form the ODE

$$\dot{\lambda}_t = -\left\{\frac{\alpha}{2} + e_t^T \Phi_t^{-1}\dot{\Phi}_t e_t\right\}\lambda_t + e_t^T \Phi_t^{-1}\frac{\widehat{\nabla^{\mathcal{I}} J_t(\theta_t)}}{\|\nabla L_t(\theta_t)\|}. \tag{5.42}$$

With γ_t (appearing in the $d\theta_t$ SDE (5.41)) satisfying,

$$\dot{\gamma}_t = -\beta(\widehat{\nabla^{\mathcal{I}} J_t(\theta_t)} - \gamma_t). \tag{5.43}$$

Remark Note that the λ_t equation (5.42) is a *linear, asymptotically stable* ODE.

The recursive solution of the Lagrangian equations (5.41), (5.42) (in itself based on (5.43) and (5.21–5.25)) forms the basic algorithm.

The control law we use is the standard CE LQ control, that is, $u_t = -K(\theta_t)x_t$. It is shown below that the parameter estimate is naturally kept bounded away from the boundary of $\mathcal{S}$ (on which K might not exist), thus making such a control law valid.

5.4   The Validity of the Lagrangian Adaptive Control Scheme

Recall that $r_t = \text{Tr}(\Phi_t)$. Then, with $\delta_t = e^{-\alpha t/2}$ and $r_t = \mathcal{O}(t)$, $\delta_t r_t \to 0$. Set $\lambda_{t_0} = 1$. The adaptive scheme consists of the Lagrangian equations (5.41, 5.42), generating $\{\theta_t, \lambda_t, t \geq t_0\}$.

Apply equations (5.41), (5.42), with initial conditions $(\theta_{t_0}, \lambda_{t_0})$ and equations (5.21)–(5.25) (for the recursive computation of $\widehat{\nabla^{\mathcal{I}}} J_t(\theta_t)$, thereby enabling the recursive computation of γ_t through (5.43)), where the control law is a standard CE LQ control, that is, $u_t = -K(\theta_t)x_t$.

Remark With the use of the *smoothed* projected gradient approximation γ_t, the formerly proposed algorithm of Levanony and Caines [32], which uses $\widehat{\nabla^{\mathcal{I}}} J_t(\theta_t)$, is considerably simplified, as the need to compute higher-order derivatives of $\widehat{\nabla^{\mathcal{I}}} J_t(\theta_t)$ (dictated by Ito's differentiation law) is avoided.

In the analysis to follow, we adopt the convention that whenever a continuous modification of a stochastic process exists, it is this modification that is considered. Furthermore, it will be assumed throughout that $\theta^* \notin \mathcal{N}$, $\mathcal{N}$ being the Lebesgue null set cited in Theorem 3.4.

Theorem 5.6 *For $t_0 > 0$, such that $\Phi_{t_0} > 0$, a.s., and arbitrary initial values $(x_{t_0}, \theta_{t_0}, \lambda_{t_0}, \gamma_{t_0}) \in \mathbb{R}^n \times \mathcal{S} \times (0, \infty) \times \mathbb{R}^{n(n+m)}$, then, with the state equation (2.1) where $(A, B) = (A^*, B^*) \in \mathcal{S}$, subjected to the feedback control $u_t = -K(\theta_t)x_t$, equations (2.1, 5.41, 5.42, 5.43) have a unique strong solution $(x, \theta, \lambda, \gamma)_t$ over $[t_0, \infty)$. Furthermore, $\{\theta_t, t \geq t_0\}$ is uniformly bounded away from $\partial\mathcal{S}$, a.s.*

Remark We note that (5.24) has a unique strong solution given by (5.10), from which it is directly derived. Next, the existence of unique strong solution to the linear ODE (5.25) is easily concluded as its dynamics matrix $F(\theta_t)$ is continuous in t.

Proof Let $z = (x, \theta, \lambda, \gamma)$ and write (2.1) and (5.41, 5.42, 5.43) in the general form

$$dz_t = F_t(z_t)dt + G_t(z_t)dw_t, \qquad (5.44)$$

where $F_t(\cdot)$ and $G_t(\cdot)$ are *random* functions of the appropriate dimensions (that is, as $\nabla L_t(\cdot)$, defined in (3.1), is a random function).

For any $n < \infty$, let $f_n \in C^\infty$, $f_n : [0, \infty) \to [0, 1]$ be such that $f_n(r) = 1, r \leq n$; $f_n(r) \in [0, 1], r \in [n, n+1]$ and $f_n(r) = 0; r \geq n+1$. Define the function

$$g_n(z, t) = f_n(\|z\|)f_n(\|\nabla L_t(\theta)\|^{-1})f_n(\lambda)f_n(1/\lambda)f_n(J(\theta)),$$

and note that g_n is jointly continuous in (z, t) (a.s.). Furthermore, let

$$F_t^n(\cdot) = F_t(\cdot)g_n(\cdot, t); \quad G_t^n(\cdot) = G_t(\cdot)g_n(\cdot, t)$$

and then F^n and G^n are globally Lipschitz in z and jointly continuous in (z, t) (a.s.). (Recall that ∇L_t is linear in θ and K is C^∞ on $\mathcal{S}$, with the latter implying that $\Psi = \Psi(\theta, x)$ is C^∞ on $\mathcal{S} \times \mathbb{R}^n$.)

Theorem 3.4.1 of Kunita [27] then implies that the SDE

$$dz_t^n = F_t^n(z_t^n)dt + G_t^n(z_t^n)dw_t, \tag{5.45}$$

with the initial condition $z_{t_0}^n = z_{t_0}g_n(z_{t_0}, t_0) \ (= z_{t_0}$ for all sufficiently large $n)$, has a unique strong solution $\{z_t^n, t \in [t_0, \infty)\}$.

Next, let $\sigma^n = \inf\{t \geq t_0 | g_n(z_t^n, t) < 1\}$, and note that (5.45) coincides with (5.44) on $[t_0, \sigma^n)$. Let $\sigma^\infty = \lim_n \sigma^n$ and define $\{z_t, t \in [t_0, \sigma^\infty)\}$ by $z_t = z_t^n, t \in [t_0, \sigma^n), n = 1, 2 \cdots$, where $z_t^l = z_t^n$, for all $1 \leq l \leq n$.

An Outline of the Steps to Follow

We now move to the proof that $\sigma^\infty = \infty$, a.s. This is done by proceeding component by component of z_t, showing that each has, with probability 1, its trajectory bounded over $[t_0, \sigma^\infty)$-uniformly bounded for $\{\theta_t, \gamma_t, x_t\}$ and bounded over finite intervals for $\{\lambda_t\}$. The uniform boundedness of $\{\gamma_t, \ t \in [t_0, \sigma^\infty)\}$ will then imply that $\{\theta_t, \ t \in [t_0, \sigma^\infty)\}$ is uniformly bounded away from $\partial\mathcal{S}$, a.s. These properties will culminate in the conclusion that $\sigma^\infty = \infty$, a.s., thereby substantiating the existence of unique strong solutions over $[t_0, \infty)$.

The Uniform Boundedness of $\{\theta_t\}$

To show that $\sup_{t_0 \leq t < \sigma^\infty} \|\theta_t\| < \infty$ a.s., we apply the Bayesian Embedding methodology. The idea is to utilize the BE to prove that *any* AML estimate is uniformly bounded and then to associate this property with the specified AML estimate constructed here.

We begin by showing that the aforementioned uniform boundedness property holds in the case of a random $\theta^*(\omega)$, distributed over $\mathbb{R}^{n(n+m)}$ with respect to a distribution which is essentially equivalent to the Lebesgue measure. Specifically, assume for a moment that θ^* is Gaussian with an a priori mean and covariance $\widehat{\theta}_0$ and $P_0 > 0$, respectively. Then, using the control $u_t = -K(\widehat{\theta}_t)x_t, \widehat{\theta}_t \triangleq E(\theta^* | \mathcal{F}_t^x)$ has the following closed form (corresponding to θ_t^{RLS}, the RLS estimate—see [8, equation (3.10)]),

$$\widehat{\theta}_t = \theta^* + [P_0^{-1} + \Phi_t]^{-1}[m_t + P_0^{-1}(\widehat{\theta}_0 - \theta^*)], \quad t \in [t_0, \sigma^\infty), \tag{5.46}$$

where $m_t = \int_0^t \Psi_s^T dw_s$. By L^1 martingale convergence, the $\mathcal{F}_t^x$-martingale $\{\widehat{\theta}_t, \mathcal{F}_t^x\}$ satisfies

$$\|\widehat{\theta}_t\| = \|E(\theta^* | \mathcal{F}_t^x)\| \leq E(\|\theta^*\| | \mathcal{F}_t^x) \to E(\|\theta^*\| | \mathcal{F}_{\sigma^\infty}^x) < \infty, \ a.s., \ \text{as } t \to \sigma^\infty.$$

Hence,

$$\sup_{t_0 \leq t < \sigma^\infty} \|\widehat{\theta}_t\| < \infty \text{ a.s.} \tag{5.47}$$

Applying the Bayesian Embedding methodology, one may immediately state that both (5.46) and (5.47) hold for the deterministic θ^* case with θ_t^{RLS} replacing the Bayesian estimate $\widehat{\theta}_t$ in both equations (outside of the Lebesgue null set cited in Theorem 3.4).

Consider a generic AML estimate θ_t: On $[t_0, \sigma^\infty)$, $\{\nabla L_t(\theta_t)\}$ is uniformly bounded since $\nabla L_t(\theta_t) \rightarrow 0$, thus, together with continuity, keeping $\{\nabla L_t(\theta_t)\}$ bounded. Furthermore, recall that by definition $\nabla L_t(\theta_t) = m_t - \Phi_t(\theta_t - \theta^*)$; hence, combined with (5.46) (with, as mentioned above, θ_t^{RLS} replacing $\widehat{\theta}_t$), one has

$$\theta_t = \theta_t^{RLS} + \Phi_t^{-1}[\theta_t^{RLS} - \theta_0^{RLS}] - \Phi_t^{-1}\nabla L_t(\theta_t), \tag{5.48}$$

which, together with the uniform boundedness of $(\nabla L_t(\theta_t), \theta_t^{RLS})$ and the monotonic decay of Φ_t^{-1}, implies that indeed $\{\theta_t, t \in [t_0, \sigma^\infty)\}$ is uniformly bounded, a.s., for all *nonrandom* θ^* outside a Lebesgue null set $\mathcal{N} \subset \mathbb{R}^{n(n+m)}$.

From a General AML Estimate to the Stochastic Lagrangian Estimate
Equations (5.30), (5.31), together with the definitions of $h_t(\theta, \lambda)$ (5.28), and $q_t(\theta)$ (5.29), as well as $\nabla L_t(\theta)$ given above (see also (3.1)), result in equations (5.41), (5.42) constructed so as to satisfy (5.30), (5.31), thereby making $\nabla L_t(\theta_t) \rightarrow 0$ by forcing $q_t(\theta_t) \rightarrow 0$ (and, simultaneously, $h_t(\theta_t, \lambda_t) \rightarrow 0$—to secure the asymptotic satisfaction of the necessary condition (5.7)), exponentially. It is therefore that the current estimate generated by equations (5.41), (5.42) is, by design, an AML estimate and hence, by the preceding result, has a uniformly bounded trajectory over $[t_0, \sigma^\infty)$, a.s.

The Estimate $\{\theta_t, \ t \in [t_0, \sigma^\infty)\}$ is Uniformly Bounded Away from $\partial\mathcal{S}$
To show that $\{\theta_t\}$ is bounded away from $\partial\mathcal{S}$, we observe that

$$\theta_t \rightarrow \theta_\tau \in \partial S \Rightarrow \nabla J(\theta_t) \rightarrow \nabla J(\theta_\tau), \ \|\nabla J(\theta_\tau)\| = +\infty \Rightarrow \tau = \sigma^\infty.$$

We note that for all $\theta \in \partial\mathcal{S}$, $\|\theta\| < \infty$, one has $J(\theta) = +\infty$, where by definition $\partial\mathcal{S}$ is a set of unstabilizable θ's. Hence, for $\theta \in \partial\mathcal{S}$, $\|\theta\| < \infty$, one also has $\|\widehat{\nabla^{\mathcal{I}}} J_t(\theta)\| = +\infty$, since by definition (5.21) $\widehat{\nabla^{\mathcal{I}}} J_t(\theta) = \mathbb{D}_t(\theta)\nabla J(\theta)$ (where $\mathbb{D}_t$ is the appropriate projection matrix made of $\widehat{D}$—see (5.21)). This, together with the definition of γ_t as a smoothed version of $\widehat{\nabla^{\mathcal{I}}} J_t(\theta_t)$ (definition (5.26)), therefore implies

$$\theta_t \rightarrow \theta_\tau \in \partial S \Rightarrow \gamma_t \rightarrow \gamma_\tau, \ \|\gamma_\tau\| = \infty, \ a.s. \tag{5.49}$$

To refute the left part of statement (5.49), namely, that $\theta_t \rightarrow \theta_\tau \in \partial\mathcal{S}$, we first show that, with no need for an artificial restriction on the evolution of λ_t, one also has that $\lambda_t \Phi_t \nabla L_t(\theta_t) \rightarrow 0$, a.e. on $\{\sigma^\infty = \infty\}$. This enables one to remove the λ-resetting mechanism, which has been employed in earlier versions of this work, see, e.g., [32]. Define

$$\rho_t = \lambda_t \Phi_t \nabla L_t(\theta_t). \tag{5.50}$$

We now derive the differential equation satisfied by $\{\rho_t\}$. With its definition (5.50), one immediately has

$$d\rho_t = \left(\dot{\lambda}_t \Phi_t \nabla L_t(\theta_t) + \lambda_t \dot{\Phi}_t \nabla L_t(\theta_t)\right) dt + \lambda_t \Phi_t d\nabla L_t(\theta_t) =$$
$$= \left(\dot{\lambda}_t \Phi_t \nabla L_t(\theta_t) + \lambda_t \dot{\Phi}_t \nabla L_t(\theta_t)\right) dt + \lambda_t \Phi_t \Psi_t^T d\tilde{v}_t - H_t d\theta_t, \tag{5.51}$$

where the second equality follows from (5.35) (recall that $H_t = \lambda_t \Phi_t^2$).

Introducing the equations for $(d\theta_t, \dot{\lambda}_t)$ (5.41, 5.42) results in an ODE of the form

$$\dot{\rho}_t = -M_t \rho_t + \xi_t, \tag{5.52}$$

with

$$M_t = \alpha I - \beta \Phi_t e_t e_t^T \Phi_t^{-1} + e_t^T \Phi_t^{-1} \dot{\Phi}_t e_t I - \Phi_t e_t e_t^T \Phi_t^{-1} \dot{\Phi}_t \Phi_t^{-1} \tag{5.53}$$

$$\xi_t = \dot{\gamma}_t + \beta h_t. \tag{5.54}$$

We note that the term driven by the pseudo-innovation $d\tilde{v}_t$ in (5.51) is canceled out by the $d\theta_t$-driven term, a fact which makes the ρ_t equation an ODE (rather than an SDE).

Consider first the case of $\sigma^\infty = \infty$.

As by Corollary 5.4 and Proposition 5.5

$$\gamma_t \to \gamma_\infty = \widehat{\nabla^{\mathcal{I}}} J_\infty(\theta_\infty) = \nabla^{\mathcal{I}} J(\theta_\infty), \tag{5.55}$$

it follows that $\dot{\gamma}_t \to 0$ (see the γ_t ODE (5.43)) and, recalling that $h_t \to 0$ (exponentially—by design), one has

$$\xi_t \to 0, \text{ a.e. on } \Sigma^\infty, \text{ as } t \to \infty, \tag{5.56}$$

with the definition of the event $\Sigma^\infty = \{\omega \in \Omega : \sigma^\infty(\omega) = \infty\}$. (We show below that $\Sigma^\infty = \Omega$.) Then, since by construction

$$h_t = \gamma_t - \rho_t \to 0, \text{ a.e. on } \Sigma^\infty, \text{ as } t \to \infty, \tag{5.57}$$

one has that

$$\rho_t \to \rho_\infty = \gamma_\infty, \text{ a.e. on } \Sigma^\infty, \text{ as } t \to \infty. \tag{5.58}$$

It can be easily verified that a.e. on Σ^∞, the matrix M_t, defined by (5.53), is dominated by the first two terms on the RHS of (5.53), thereby making it positive definite for all large enough t. This is due to (i) e_t is a unit vector, (ii) $\Phi_t \uparrow \infty$, (iii)

$\Phi_t^{-1}\dot{\Phi}_t \rightarrow 0$, and (iv) $\beta\Phi_t e_t e_t^T \Phi_t^{-1} \leq \beta I$ for all t. One therefore concludes that a.e. on Σ^∞,

$$\lim\inf_{t\rightarrow\infty}\lambda_{min}(M_t) \geq \alpha - \beta > 0. \tag{5.59}$$

Then, a.e. on Σ^∞: Since $\xi_t \rightarrow 0$ (5.56) and, as clearly $\lambda_{min}(M_t) \geq c = (\alpha-\beta)/2 > 0$, for all $t \geq \tau_c$ for some $\tau_c = \tau_c(\omega) < \infty$, one observes that the ξ_t-perturbed linear ODE (5.52) satisfies the conditions of Khalil [22, Lemma 5.4], thereby making (5.52) locally input-to-state stable [22, Definition 5.2] for all initial times $t_0 \geq \tau_c$. Specifically: (1) The origin is globally, $[\tau_c, \infty)$-uniformly asymptotically stable equilibrium for the unforced ($\xi_t \equiv 0$) ODE (5.52) and (2) $\xi_t \rightarrow 0$. (Note that here, t_0 refers to the notation given in [22, Definition 5.2], not to be confused with the initial time t_0 stated in this work.) Then, using the fact that $\xi_t \rightarrow 0$, as $t \rightarrow \infty$, a.s., one may take $t = 2t_0$ in [22, Definition 5.2] to directly conclude that

$$\rho_t \rightarrow 0, \ as \ t \rightarrow \infty, \ a.e. \ on \ \Sigma^\infty. \tag{5.60}$$

Hence, with continuity, one has

$$\sup_{t_0 \leq t < \sigma^\infty} \|\rho_t\| < \infty, \ a.s., \tag{5.61}$$

where we note that (5.61) holds also in case $\sigma^\infty < \infty$, as the solution of the linear ODE (5.52) with bounded coefficients and a diminishing input has no *finite* escape time.

Now, given (5.61), together with the *designed* exponential decay of h_t, one has that $\gamma_t = h_t + \rho_t$ is uniformly bounded over $[t_0, \sigma^\infty)$, which in turn implies that θ_t is *uniformly* bounded away from ∂S on $[t_0, \sigma^\infty)$.

The Boundedness of $\{\lambda_t\}$ over Finite Intervals
By the definition of $q_t(\theta_t)$ (5.29), together with its forced exponential decay (5.31) and, by choosing δ_{t_0} such that $q_{t_0}(\theta_{t_0}) > 0$, one has

$$\|\nabla L_t(\theta_t)\|^2 = 2q_{t_0}(\theta_{t_0})e^{-\alpha(t-t_0)} + \delta_t^2 > 0, \ \forall t < \infty. \tag{5.62}$$

Note that nominator of the input to (5.42), $\widehat{\nabla}J_t(\theta_t) = \mathcal{O}(\|\gamma_t\|)$, hence is uniformly bounded. This, together with (5.62), implies that the linear ODE (5.42), generating $\{\lambda_t, \ t_0 \leq t < \sigma^\infty\}$, is perturbed by a bounded input over *finite* intervals and has bounded coefficients (where, with $\Phi_t^{-1}\dot{\Phi}_t \rightarrow 0$, becoming, as $t \rightarrow \infty$, an asymptotically stable linear ODE), and hence its solution is a.s. bounded over finite intervals within $[t_0, \sigma^\infty)$, that is, it does not have a finite escape time, yet may diverge as $t \rightarrow \infty$ (on $\Sigma^\infty = \{\sigma^\infty = \infty\}$).

The Uniform Boundedness of $\{x_t\}$

Summarizing the above, one has that, with probability 1, (a) $\{\theta_t, \ t \in [t_0, \sigma^\infty)\}$ uniformly bounded (by virtue of the Bayesian Embedding argument), (b) $\{\gamma_t, \ t \in [t_0, \sigma^\infty)\}$ uniformly bounded, and (c) $\{\lambda_t, t \in [t_0, \sigma^\infty)\}$ bounded over finite intervals. Then, with λ_t not explicitly appearing in (2.1)-the SDE generating the state $\{x_t\}$ (where $K_t = K(\theta_t)$) – and given (a) and (b), one may invoke Proposition 6.1 ([9, Theorem 2.1, Part (i)] for $\sigma^\infty < \infty$ and [9, Theorem 2.1, Part (ii)] for $\sigma^\infty = \infty$), to state that (d) $\sup_{t \geq t_0} \|x_t\| < \infty$, finally making, together with (a), (b) and (c), (e) $\sigma^\infty = \infty \Rightarrow \Sigma^\infty = \Omega$. $\qquad\square$

5.5 From Convergence to Consistency to Optimal Performance

Corollary 5.7 *Under the conditions of Theorem 5.6,*

$$\theta_t \to \theta_\infty(\omega) \in \mathcal{I} \quad as\ t \to \infty,\ a.s.$$

Proof As a unique strong solution exists on $[t_0, \infty)$, it is necessary that by (5.29, 5.31) and the choice of $\{\delta_t\}$, that

$$\|\nabla L_t(\theta_t)\| \to 0 \quad \text{a.s. } t \to \infty \text{ a.s.}$$

Hence, θ_t is an $\mathcal{AML}$ estimate and by Theorem 3.4 converges to a finite limit in $\mathcal{I}$. $\qquad\square$

Theorem 5.8 *Suppose that $\Phi_{t_0} > 0$. Then with probability 1, $\theta_\infty(\omega) = \theta^*$, that is, the estimate θ_t is strongly consistent, and the associated CE-based adaptive scheme is asymptotically optimal in the sense of (2.8).*

Proof First note that, with $\sigma^\infty = \infty$ a.s., then with (5.58) and (5.60), one obtains

$$\lim_{t\to\infty}\gamma_t = 0,\ a.s. \tag{5.63}$$

Next, Corollary 5.7, which gives $\theta_t \to \theta_\infty \in \mathcal{I}$, a.s., as $t \to \infty$, and Proposition 5.5, giving $\gamma_t \to \nabla^{\mathcal{I}} J(\theta_\infty)$, a.s., as $t \to \infty$, combined with (5.63), together lead to

$$\lim_{t\to\infty}\gamma_t = \nabla^{\mathcal{I}} J(\theta_\infty) = 0,\ a.s. \tag{5.64}$$

Finally, by (4.21) and Lemma 4.1, we obtain

$$\theta_t \to \theta_\infty \in \{\theta \in \mathcal{I} | \nabla^{\mathcal{I}} J(\theta) = 0\} = \{\theta^*\},\ a.s.,\ as\ t \to \infty. \tag{5.65}$$

Optimality follows from Theorem 5.1. $\qquad\square$

We end with the following Lemma which establishes that Φ_t is positive definite for all $t > 0$.

Lemma 5.9 *Given the adaptive control scheme summarized by (5.21–5.25) and (5.41–5.43), one has*

$$\Phi_t > 0, \ \forall t > 0, \ a.s. \tag{5.66}$$

Proof Since by definition $\Phi_t = \int_0^t \Psi_s^T \Psi_s ds = BlockDiag^{(n)}\{Q_t\}$, with $Q_t = \int_0^t \varphi_s \varphi_s^T ds$, $\varphi_s = (x_s^T, u_s^T)$, it suffices to show that $Q_t > 0$ for all $t > 0$, a.s. where, with the control signal in the feedback form $u_t = -K(\theta_t)x_t, \ t \geq 0$, one has

$$Q_t = \begin{bmatrix} \int_0^t x_s x_s^T ds & -\int_0^t x_s x_s^T K^T(\theta_s)ds \\ -\int_0^t K(\theta_s)x_s x_s^T ds & \int_0^t K(\theta_s)x_s x_s^T K_s^T(\theta_s)ds \end{bmatrix}. \tag{5.67}$$

Suppose that for some $t > 0$ there exists an event $\mathcal{A}_t \in \mathcal{F}_t^x$ of nonzero measure such that Q_t is singular, a.e. on $\mathcal{A}_t$. Then there exists an $\mathbb{R}^{n+m}$ vector $a \neq 0$, $a \in \mathcal{F}_t^x$ such that $a^T Q_t a = 0$, a.e. on $\mathcal{A}_t$.
Set $K_s = K(\theta_s)$, $s \in [0, t]$ and write $a = (a_1^T, a_2^T)^T$ with $a_1 \in \mathbb{R}^n$, $a_2 \in \mathbb{R}^m$. Then, a.e. on $\mathcal{A}_t$,

$$a^T Q_t a = \int_0^t a^T \begin{bmatrix} dP_s & -dP_s K_s^T \\ -K_s dP_s & K_s dP_s K_s^T \end{bmatrix} a = 0. \tag{5.68}$$

But since $P_t = \int_0^t x_s x_s^T ds$ is strictly increasing, the $\mathbb{R}^1$ valued integral in (5.68) is increasing in $t > 0$, and hence its *integrand* has value 0 for all s, $0 \leq s \leq t$, a.e. on $\mathcal{A}_t$.
In light of the specified quadratic form of (5.68), it follows that

$$[a_1^T - a_2^T K_s]x_s = 0 \ a.e. \ on \ \mathcal{A}_t, \ 0 \leq s \leq t. \tag{5.69}$$

Let $\mathbf{1}_{\mathcal{A}_t}$ denote the indicator function of $\mathcal{A}_t$. Then (5.69) is equivalent to

$$\mathbf{1}_{\mathcal{A}_t} a_1^T x_s = \mathbf{1}_{\mathcal{A}_t} a_2^T K_s x_s, \ a.s., \ 0 \leq s \leq t. \tag{5.70}$$

But this yields a contradiction, since, with $a \neq 0$ eliminating the trivial solution, equating the *nonzero* quadratic variations of the expressions on the left and right sides of (5.70), while recalling that with $C = I$, $< x >_s = Is$, gives

$$\mathbf{1}_{\mathcal{A}_t} \|a_1\|^2 s = \mathbf{1}_{\mathcal{A}_t} \|K_s a_2\|^2 s, \ \forall s \in [0, t], \ a.s. \tag{5.71}$$

So, by removing s from both sides of (5.71), one has that

$$\mathbf{1}_{\mathcal{A}_t} \|a_1\| = \mathbf{1}_{\mathcal{A}_t} \|K_s a_2\|, \ \forall s \in [0, t], \ a.s., \tag{5.72}$$

thus making the *constant* LHS equal to the *nonconstant* RHS $\forall s \in [0, t]$, obviously establishing a contradiction. $\qquad\square$

5.6   A Discussion of the Adaptive Scheme's Initialization Time t_0

Our analysis is based on the understanding that the adaptive scheme is initialized at some $t_0 > 0$, where, at t_0, one has $P_{t_0} \geq \varepsilon I$, $\Phi_{t_0} \geq \varepsilon I$, with I being the identity matrix of the appropriate dimension, $\varepsilon > 0$, an arbitrary small constant. Given the above, one may define t_0 as a suitable stopping time.

This leaves one to decide about the control policy over $[0, t_0]$. To that end, we examine the following three options:

(1) One may use any erratic feedback control $u_t = -K(\theta_t)x_t$, $t \in [0, t_0]$, with $\theta_t \in \mathcal{S}$, so as to excite the state x_t and the regression process $\varphi_t = [x_t^T, u_t^T]^T$, so that indeed, t_0 will be small. Specifically, one may choose to apply over $[0, t_0]$ the RLS scheme proposed by Caines [5] and Caines and Levanony [7], which carries a projection mechanism that is to reset the estimate θ_t, to a preassigned $\theta_0 \in \mathcal{S}$, once θ_t leaves a certain compact subspace of $\mathcal{S}$.

(2) Another, much simpler, straightforward way would leave $\theta_t = \theta_0 \in \mathcal{S}$, $0 \leq t \leq t_0$. This, however, will result in $t_0 = \infty$, as the state x_t and the control $u_t = -K(\theta_0)x_t$ become linearly dependent, hence, ruling out a positive-definite Φ_t, $\forall t \geq 0$.

(3) On the other hand, one may wish to commence the Lagrangian adaptation scheme at $t_0 = 0$. This is however impractical: In this case one will *artificially* set $P_0 = \varepsilon I$, $\Phi_0 = \varepsilon I$, and then (will try to) continue as above. However, the fact that by definition $\nabla L_0(\theta) = 0$, with no reasonable, artificially set nonzero initial condition as justification, makes such a prospect unrealizable, in terms involving $1/\|\nabla L_t(\theta_t)\|$ play a key role in the algorithm, in particular in the differential equations generating $\{\theta_t, \lambda_t\}$. See, e.g., (5.85, 5.86) below. Moreover, putting such artificial nonzero initial condition, say $\Phi_0 = \varepsilon I$, will render the desired exponential decay of $\nabla L_t(\theta_t)$ out of reach. We therefore conclude in adopting option (1) above.

5.7   A Summary of the Lagrangian Adaptation Scheme

For completeness, we rewrite the Lagrangian adaptation algorithm equations.
   The controlled state $\{x_t\}$ is generated by the SDE

$$dx_t = A^* x_t dt + B^* u_t dt + dw_t, \quad x_0, \qquad (5.73)$$

with the control taking the feedback form

$$u_t = -K(\theta_t)x_t. \qquad (5.74)$$

(i) *Computation of the feedback matrix K*

For $\theta_t = col(A_t, B_t)$, solve for V to compute K:

$$A_t^T V + V A_t - V B_t B_t^T V + I = 0, \tag{5.75}$$

$$K = K(\theta_t) = B_t^T V(\theta_t). \tag{5.76}$$

(ii) *Construction of the smoothed projected gradient of the LQ cost*

Generate $\{\Pi_t(i, j), \ t \geq t_0\}$, $\{\zeta_t(i, j), \ t \geq 0\}$, $i = 1, 2 \cdot \cdot n$, $j = 1, 2 \cdots m$, solutions to

$$d\Pi_t(i, j)$$

$$= P_t^{-1}\{\zeta_t(i, j)dx_t^T + x_t \zeta_t^T(i, j)F^T(\theta_t)dt - [\zeta_t(i, j)x_t^T + x_t \zeta_t^T(i, j)]\Gamma_t dt$$

$$+ \int_0^t (\zeta_s(i, j)x_s^T + x_s \zeta_s^T(i, j))ds\, P_t^{-1} x_t [x_t^T \Gamma_t dt - dx_t^T]\}, \ t \geq t_0, \ \Pi_{t_0}(i, j), \tag{5.77}$$

$$\dot{\zeta}_t(i, j)$$

$$= F(\theta_t)\zeta_t(i, j) + \Pi_t^T(i, j)x_t, \ \zeta_0(i, j) = 0, \ t \geq 0, \tag{5.78}$$

$$\text{with } \Pi_t(i, j) = \frac{\partial F^T(\theta_t)}{\partial B_{ij}}, \ 0 \leq t < t_0.$$

The initial condition of (5.77), namely, $\Pi_{t_0}(i, j)$, is computed by (5.10) with $t = t_0$. This, with the terms of (5.10), computed recursively in an obvious way (e.g., $dP_t = x_t x_t^T dt$, etc.). Also used are $F(\theta) = A - BK(\theta)$, $\theta = col(A, B)^T$, $P_t = \int_0^t x_s x_s^T ds$, $\Gamma_t = P_t^{-1} \int_0^t x_s dx_s^T$. The gain matrix $\Pi_t(i, j)$, $0 \leq t < t_0$ in (5.78) is computed through

$$\Pi_t(i, j) = \frac{\partial F^T(\theta_t)}{\partial B_{ij}}$$

$$= -([\frac{dB}{dB_{ij}}B_t^T + B_t \frac{dB^T}{dB_{ij}}]V(\theta_t) + B_t B_t^T \frac{\partial V(\theta_t)}{\partial B_{ij}})^T, \ 0 \leq t < t_0, \tag{5.79}$$

since $dA/dB = 0$. (See the full expression of $\Pi_t(i, j)$ given by (5.82) below.) The partial derivative $\partial V/\partial B_{ij}$ in the last term on the RHS of (5.79) is the solution to

$$F^T \frac{\partial V}{\partial B_{ij}} + \frac{\partial V}{\partial B_{ij}}F - V[\frac{dB}{dB_i j}B^T + B\frac{dB^T}{dB_{ij}}]V = 0. \tag{5.80}$$

Then, for $t \geq t_0$, run the $\zeta_t(i, j)$ ODE (5.78), with $\Pi_t(i, j)$ taken to be the solution generated by the SDE (5.77). (We note the discontinuity at t_0 in the input to (5.78).)

Utilize Π_t to compute the (estimated) projected gradient $\widehat{\nabla^{\mathcal{I}}} J$: With $J(\theta) = \mathrm{Tr} V(\theta)$,

$$\widehat{\nabla^{\mathcal{I}}} J_t(\theta) = \begin{bmatrix} \widehat{D}_t(\theta) \\ I \end{bmatrix} [\widehat{D}_t^T(\theta) \widehat{D}_t(\theta) + I]^{-1} [\widehat{D}_t^T(\theta) \frac{\partial J(\theta)}{\partial col A} + \frac{\partial J(\theta)}{\partial col B}], \quad \theta = \theta_t,$$

(5.81)

where the entries of $\widehat{D}_t(\theta)$ are given by $\{X_{ij}(k, \ell)\}$, which are the solutions to the quantities

$$\Pi_t(i, j) = \left(X_{ij} - \left[\frac{dB}{dB_{ij}} B^T + B \frac{dB^T}{dB_{ij}} \right] V(\theta) - B B^T \frac{dV(\theta)}{dB_{ij}} \right)^T \quad (5.82)$$

$$\frac{dV_{pq}(\theta)}{dB_{ij}} = \sum_{k=1}^{n} \sum_{\ell=1}^{n} \frac{\partial V_{pq}}{\partial A_{k\ell}} X_{ij}(k, \ell) + \frac{\partial V_{pq}(\theta)}{\partial B_{ij}}, \quad \theta = \theta_t, \quad (5.83)$$

where $i = 1, 2 \cdots n$, $j = 1, 2 \cdots m$, $p, q = 1, 2 \cdots n$. (Again, note that, for $\theta \in \mathcal{I}$, $\Pi_t(i, j) = dF^T(\theta)/dB_{ij}$, one has that $X_{ij}(k, \ell) = dA_{k\ell}/dB_{ij}$.)

The smoothed version γ_t of the (approximated) projected gradient evolves according to

$$\dot{\gamma}_t = \beta(\widehat{\nabla^{\mathcal{I}}} J_t(\theta_t) - \gamma_t), \quad \gamma_0 = 0. \quad (5.84)$$

(iii) *Parameter estimation*

Recall the definitions of Ψ_t (2.3), Φ_t (3.2), $\nabla L_t(\theta)$ (3.1), $H_t = \lambda_t \Phi_t^2$, $\ell_t(\theta) = \Phi_t \nabla L_t(\theta)$, $e_t = \nabla L_t(\theta_t)/\|\nabla L_t(\theta_t)\|$, $\mu_t = \lambda_t/\|\nabla L_t(\theta_t)\|$, $\delta_t = \exp\{-\alpha t/2\}$, γ_t (5.26), $h_t(\theta, \lambda)$ (5.28), and $q_t(\theta)$ (5.29). Fix some $\alpha > \beta > 0$. Choose an arbitrary $\theta_{t_0} \in \mathcal{S}$.

The parameter estimation algorithm takes the form

$$d\theta_t = \Phi_t^{-1} \Psi_t^T \{dx_t - \Psi_t \theta_t dt\} + \mu_t H_t^{-1} \ell_t(\theta_t) \{\alpha q_t dt - \delta_t d\delta_t\}$$

$$+ H_t^{-1} [I - \mu_t \ell_t(\theta_t) \ell_t^T(\theta_t) H_t^{-1}]$$

$$\times \left\{ \lambda_t \dot{\Phi}_t \nabla L_t(\theta_t) - \beta(\widehat{\nabla^{\mathcal{I}}} J_t(\theta_t) + \lambda_t \Phi_t \nabla L_t(\theta_t)) \right\} dt,$$

$$\theta_{t_0} \in \mathcal{S}, \quad (5.85)$$

with the adjoint variable λ_t solved through the ODE

$$\dot{\lambda}_t = -\left\{ \frac{\alpha}{2} + e_t^T \Phi_t^{-1} \dot{\Phi}_t e_t \right\} \lambda_t + e_t^T \Phi_t^{-1} \frac{\widehat{\nabla^{\mathcal{I}}} J_t(\theta_t)}{\|\nabla L_t(\theta_t)\|}, \quad \lambda_{t_0} = 1. \quad (5.86)$$

5.8 Computation of the Regret Rate

Following recent similar works on the problem under study, see, e.g., [17, 39] and references therein, we look at the rate at which the t-normalized cumulative quadratic cost approaches the optimal $J(\theta^*)$ cost. Let

$$Regret(t) = \frac{1}{t} \int_0^t (\|x_s\|^2 + \|u_s\|^2) ds - J(\theta^*). \tag{5.87}$$

Then, by [8, Equation (5.57)], while noting that in that equation $J(\theta^*) = \mathrm{Tr}V$, one may easily compute the regret rate of the Lagrangian adaptive control scheme presented here, thereby recovering the well-known result that $Regret(t) = \mathcal{O}(1/\sqrt{t})$. This is specified in the following:

By [8, Equation (5.57)], which gives the t-normalized cumulative quadratic cost (with $J(\theta^*) = \mathrm{Tr}V$, $V = V(\theta^*)$), one has

$$Regret(t) = \frac{1}{t}M_t + \frac{1}{t}(x_0^T V x_0 - x_t^T V x_t) + \frac{1}{t} \int_0^t x_s^T (K^* - K_s)^T (K^* - K_s) x_s ds, \tag{5.88}$$

where $K^* = K(\theta^*)$, $K_s = K(\theta_s)$, $M_t = 2 \int_0^t x_s^T V dw_s$. Assume that

$$K_t - K^* = \mathcal{O}(\|\theta_t - \theta^*\|), \tag{5.89}$$

a conjecture we show below to hold.

Then, with (3.1) together with the forced exponential decay of $\nabla L_t(\theta_t)$, giving

$$\frac{1}{t}\nabla L_t(\theta_t) = \frac{1}{t}m_t - \frac{\Phi_t}{t}(\theta_t - \theta^*) = \mathcal{O}(\frac{e^{-\alpha t}}{t}), \tag{5.90}$$

and recalling that $\{m_t = \int_0^t \Psi_s^T dw_s, t \geq 0\}$ is a martingale with $< m >_t = \Phi_t = \mathcal{O}(t)$, one has

$$\|\theta_t - \theta^*\| = \mathcal{O}(\frac{\|m_t\|}{t}) = \mathcal{O}(1/\sqrt{t}). \tag{5.91}$$

Relation (5.91), together with conjecture (5.89), gives

$$\frac{1}{t} \int_0^t x_s^T (K^* - K_s)^T (K^* - K_s) x_s ds = \mathcal{O}(\frac{1}{t} \int_0^t \|K^* - K_s\|^2 ds)$$

$$= \mathcal{O}(\frac{1}{t} \int_\varepsilon^t \frac{1}{s} ds) = \mathcal{O}(\frac{log(t)}{t}), \tag{5.92}$$

where the first equality rests on the fact that $\sup_{t \geq 0} \|x_t\| < \infty$, $a.s.$—see Proposition 6.1.

Summarizing the terms on the RHS of (5.88) while noting that $\{M_t, \ t \geq 0\}$ is a martingale with $< M >_t = \mathcal{O}(t)$, hence giving $M_t/t = \mathcal{O}(1/\sqrt{t})$, one obtains

$$Regret(t) = \mathcal{O}(1/\sqrt{t}). \tag{5.93}$$

It remains to prove (5.89). As has been shown in detail in Chap. 3 above, the estimation errors of $\theta_t = col(A_t, B_t)$ and of $\alpha_t = col\{A^* - B^* K(\theta_t)\}$ are closely associated with their least squares counterparts. Specifically, in the case of the estimation of the closed-loop dynamics $\alpha_t^* = col\{A^* - B^* K(\theta_t)\}$, one may deduce the convergence rate of α_t from that of its LS counterpart—see the Proof of Theorem 3.4. We utilize this property as follows.

Rewrite the state equation (2.1) (with $A = A^*$, $B = B^*$) as (3.19) above,

$$dx_t = X_t \alpha_t^* dt + dw_t, \quad x_0, \tag{5.94}$$

where X_t is the $n \times n^2$ matrix,

$$X_t = \begin{bmatrix} x_t^T & 0 & \cdots & 0 \\ 0 & x_t^T & 0 & 0 \\ 0 & \cdots & \cdots & x_t^T \end{bmatrix}. \tag{5.95}$$

A standard RLS estimate of α_t^*, based on the observed state x_t, the solution to (5.94), evolves according to

$$d\alpha_t^{RLS} = R_t^{-1} X_t^T [dx_t - X_t \alpha_t^{RLS} dt], \tag{5.96}$$

with

$$R_t = \int_0^t X_s^T X_s ds + R_0, \quad R_0 > 0. \tag{5.97}$$

The RLS estimation error then takes the closed form

$$\alpha_t^{RLS} - \alpha_t^* = R_t^{-1} [\int_0^t X_s^T dw_s + R_0 (\alpha_0^{RLS} - \alpha_0^*)]. \tag{5.98}$$

Then, the fact that $\int_0^t x_s x_s^T ds = \mathcal{O}(t)$, together with a martingale LLN and (5.98), gives

$$\alpha_t^{RLS} - \alpha_t^* = \mathcal{O}(1/\sqrt{t}), \tag{5.99}$$

thus enabling us to state that

$$\alpha_t - \alpha_t^* = \mathcal{O}(1/\sqrt{t}). \tag{5.100}$$

Given the above, then with $K_t = K(\theta_t)$, $K^* = K(\theta^*)$, one has the closed-loop estimation error,

$$A_t - B_t K_t - (A^* - B^* K^*) = A_t - A^* - (B_t K_t - B^* K^*)$$
$$= A_t - A^* - (B_t - B^*)K^* + B_t(K_t - K^*). \tag{5.101}$$

Hence, substantiating (5.89), which with (5.91), namely, $\theta_t - \theta^* = \mathcal{O}(1/\sqrt{t})$, gives

$$K_t - K^* = \mathcal{O}(1/\sqrt{t}). \tag{5.102}$$

Chapter 6
Proof of Theorem 5.2

Abstract A key feature of our methodology is the consistent approximation of the derivative with respect to the system input matrix parameters of the time-varying, closed-loop dynamics matrix process; that consistency is rigorously proven in this chapter by the use of stochastic Picard iterations.

Keywords Generalized stochastic Picard iterations · Hurwitz · Ito's differentiation rule · Martingale LLN · Locally input-to-state stable · Transition matrix process · Cauchy sequence · Uniform convergence

6.1 Proof Methodology

The proof involves successive consistent approximations of $dF^T(\theta_\infty)/dB$. The idea is to construct Picard-type iterations $\{\Pi_t^{(n)}, \zeta_t^{(n)}, t \geq 0\}$, $n = 0, 1, 2 \cdots$, of the pair $\{\Pi_t, \zeta_t, t \geq 0\}$, defined by (5.10), (5.11) (respectively), so that each iteration will satisfy (5.12), namely,

$$\Pi_t^{(n)}(i, j) \to dF^T(\theta_\infty)/dB_{ij}, \quad \text{as } t \to \infty, \ a.s. \tag{6.1}$$

With the first $n = 0, 1$ approximations shown below to satisfy (6.1), we construct the general Picard iterations for $n \geq 2$. Utilizing the properties established for $n = 0, 1$, we explicitly show that successive iteration differences diminish to zero, as $n \to \infty$, geometrically, *uniformly* on $[0, \infty]$. This will make $\{\Pi_t^{(n)}, \zeta_t^{(n)}, t \geq 0, n \geq 2\}$ a Cauchy sequence in $\mathcal{C}[0, \infty)$ (equipped with the sup-norm), and hence a convergent sequence, with a limit given by $\{\Pi_t, \zeta_t, t \geq 0\}$, satisfying, by construction, (5.10, 5.11). With this convergent property established, it is shown that (6.1) holds for $\{\Pi_t, t \geq 0\}$.

6.2 Preliminary Results

We begin with the several preliminary results. Recall that the state $\{x_t, \ t \geq 0\}$ evolves according to

$$dx_t = F^*(\theta_t)x_t dt + C dw_t, \ x_0, \tag{6.2}$$

where $F^*(\theta_t) = A^* - B^* K(\theta_t)$. By Theorem 3.4, one has that

$$F^*(\theta_t) \to F^*(\theta_\infty) = F(\theta_\infty) = A_\infty - B_\infty K(\theta_\infty), \ as \ t \to \infty, \ a.s., \tag{6.3}$$

with the first equality due to the characterization of $\mathcal{I}$—see (3.52). Moreover, the fact that $F(\theta_t)$ is asymptotically stable for all $t \geq 0$ makes the limit $F^*(\theta_\infty) = F(\theta_\infty)$ Hurwitz (asymptotically stable). That Hurwitz limit property enables us to state the following.

Proposition 6.1 ([9, Theorem 2.1]) *For the solution to (6.2), with $F^*(\theta_t) \to F^*(\theta_\infty)$ Hurwitz (a.s.), it holds that*

$$\sup_{t \geq 0} \|x_t\| < \infty, \ a.s. \tag{6.4}$$

Next, we verify the following result which, while appearing to be standard, could not found in the LTI SDEs literature.

Theorem 6.2 *Let $\{x_t, \ t \geq 0\}$ be the solution to (6.2). Then, under (6.3), with $P_t = \int_0^t x_s x_s^T ds$ and $CC^T > 0$, one has*

$$\lim_{t \to \infty} P_t / t = \int_0^\infty exp[F(\theta_\infty)t]CC^T exp[F^T(\theta_\infty)t]dt > 0, \ a.s. \tag{6.5}$$

Proof The use of Ito's differentiation rule, together with the x_t-state equation (6.2), gives

$$x_t x_t^T - x_0 x_0^T$$
$$= \int_0^t x_s x_s^T ds F^T(\theta_\infty) + F(\theta_\infty) \int_0^t x_s x_s^T ds + CC^T t + M(t)$$
$$+ \int_0^t x_s x_s^T \left(F^*(\theta_s) - F(\theta_\infty)\right)^T ds + \int_0^t \left(F^*(\theta_s) - F(\theta_\infty)\right) x_s x_s^T ds, \tag{6.6}$$

where $M(t) = \int_0^t (x_s dw_s^T C^T + C dw_s x_s^T)$, $t \geq 0$, is a martingale, which, due to the martingale LLN, satisfies $M(t)/t \to 0$ as $t \to \infty$, a.s. This follows from the fact that, by (6.4), its quadratic variation matrix process is of $\mathcal{O}(t)$, see also

[8, Lemma 5.5], giving

$$\limsup_{t \to \infty} \mathrm{Tr} P_t / t < \infty, \quad a.s. \tag{6.7}$$

Furthermore,

$$\limsup_{t \to \infty} \frac{1}{t} \int_0^t x_s x_s^T \left(F^*(\theta_s) - F(\theta_\infty) \right)^T ds$$

$$\leq \limsup_{t \to \infty} \frac{1}{t} \int_r^t x_s x_s^T \, ds \, \sup_{s \geq r} \| F^*(\theta_s) - F(\theta_\infty) \|, \tag{6.8}$$

which holds for all $r > 0$. With (6.3) and (6.7), one may take $r \to \infty$ to conclude that

$$\lim_{t \to \infty} \frac{1}{t} \int_0^t x_s x_s^T \left(F^*(\theta_s) - F(\theta_\infty) \right)^T ds = 0, \quad a.s. \tag{6.9}$$

Next, as by (6.4) $\sup_{t \geq 0} \| x_t \| < \infty$, a.s., it follows that $P = \lim_{t \to \infty} P_t / t$ exists: This is shown by dividing (6.6) by t and taking the limit as $t \to \infty$, which gives

$$0 = P F^T(\theta_\infty) + F(\theta_\infty) P + C C^T. \tag{6.10}$$

Finally, one may conclude, see, e.g., Khalil [22, Theorem 3.6 and Equation (3.13)], that P, the unique solution to (6.10), is indeed given by the RHS of (6.5).

Recall the definitions $P_t = \int_0^t x_s x_s^T \, ds$ and $\Gamma_t = P_t^{-1} \int_0^t x_s dx_s^T$. Last among the preliminary results is the following.

Proposition 6.3

$$\lim_{t \to \infty} \Gamma_t = F^T(\theta_\infty), \quad a.s. \tag{6.11}$$

Proof To verify (6.11), write

$$\Gamma_t = P_t^{-1} \int_0^t x_s dx_s^T = P_t^{-1} \int_0^t x_s x_s^T F^{*T}(\theta_s) ds + P_t^{-1} \int_0^t x_s dw_s^T C^T$$

$$= P_t^{-1} \int_0^t x_s x_s^T \left(F^*(\theta_s) - F(\theta_\infty) \right)^T ds + P_t^{-1} \int_0^t x_s dws^T C^T + F^T(\theta_\infty). \tag{6.12}$$

With (6.5), a martingale LLN yields

$$\lim_{t \to \infty} P_t^{-1} \int_0^t x_s dw_s^T C^T = 0, \quad a.s. \tag{6.13}$$

Furthermore,

$$\lim_{t \to \infty} P_t^{-1} \int_0^t x_s x_s^T \left(F^*(\theta_s) - F(\theta_\infty) \right)^T ds$$

$$\leq \lim_{t \to \infty} P_t^{-1} \Big\{ \int_0^r x_s x_s^T \left(F^*(\theta_s) - F(\theta_\infty) \right)^T ds$$

$$+ \int_r^t x_s x_s^T ds \big[\sup_{s \geq r} \| F^*(\theta_s) - F(\theta_\infty) \| \big] \Big\}$$

$$= \lim_{t \to \infty} P_t^{-1} \int_r^t x_s x_s^T ds \big[\sup_{s \geq r} \| F^*(\theta_s) - F(\theta_\infty) \| \big]$$

$$\leq I \limsup_{s \geq r} \| F^*(\theta_s) - F(\theta_\infty) \|. \tag{6.14}$$

By Theorem 3.5, $\theta_t \to \theta_\infty \in \mathcal{I}$, which, together with the definition of $\mathcal{I}$ making $F^*(\theta_\infty) = F(\theta_\infty)$, one has that $F^*(\theta_t) \to F^*(\theta_\infty) = F(\theta_\infty)$, as $t \to \infty$, a.s. Taking $r \to \infty$ leads to

$$\lim_{t \to \infty} P_t^{-1} \int_0^t x_s x_s^T \left(F^*(\theta_s) - F(\theta_\infty) \right)^T ds = 0, \quad a.s., \tag{6.15}$$

which, together with (6.13), substantiates (6.11).   □

We now present the first two approximation of the matrix processes $\{\Pi_t(i, j)\}, i = 1, 2 \cdots n; j = 1, 2 \cdots m$ and validate their consistency in the estimation of $dF^T(\theta_\infty)/dB_{ij}$.

6.3   A Zero-Order Approximation of Π

For $F(\theta) = A - BK(\theta)$, set

$$\Pi_t^{(0)}(i, j) = \frac{d}{dB_{ij}} F^T(\theta_t), \quad i = 1, 2 \cdots n; j = 1, 2 \cdots m. \tag{6.16}$$

Then, given that F is continuously differentiable together with the a.s. convergence $\theta_t \to \theta_\infty \in \mathcal{I}$, one concludes that

$$\Pi_t^{(0)}(i, j) \to \frac{d}{dB_{ij}} F^T(\theta_\infty), \quad \text{as } t \to \infty \text{ a.s.}, \quad i = 1, 2 \cdots n; j = 1, 2 \cdots m \tag{6.17}$$

Remark At this point it is worth recalling the Remark made below (5.12): The (full) differentiation of $F(\theta)$ with respect to B inherently involves the computation of dA/dB, which is impossible to evaluate for all $t \geq t_0$, as the functional dependence of A on B is unknown throughout $[t_0, \infty)$. Moreover, the functional dependence of A on B only occurs at the asymptotic limit, as $\theta_t \to \theta_\infty \in \mathcal{I}$, with that function, depending upon θ^*, is determined by the definition of $\mathcal{I}$. It is therefore that the zero-order approximation described above is set for the purpose of the analysis to follow which culminates in a purely data-based approximation.

6.4 A First-Order Approximation of Π

The next approximation of $dF^T(\theta_\infty)/dB_{ij}$ is constructed with $\zeta_t^{(1)}(i, j)$ built so as to approximate the derivative of x_t with respect to B_{ij} (itself an approximation of $\zeta_t(i, j)$, the solution to (5.11)). The $\{\Pi_t^{(1)}(i, j),\ \zeta_t^{(1)}(i, j),\ t \geq 0\}$ approximations are given by:

$$\Pi_t^{(1)}(i, j) = P_t^{-1}\Big\{ \int_0^t \zeta_s^{(1)}(i, j)dx_s^T + \int_0^t x_s d\zeta_s^{(1)T}(i, j)$$

$$- \int_0^t (\zeta_s^{(1)}(i, j)x_s^T + x_s \zeta_s^{(1)T}(i, j))ds\,\Gamma_t\Big\}, \qquad (6.18)$$

where $\{\zeta_t^{(1)}(i, j),\ t \geq 0\}$ is defined via the ODE

$$\dot{\zeta}_t^{(1)}(i, j) = F(\theta_t)\zeta_t^{(1)}(i, j) + \Pi_t^{(0)}(i, j)x_t, \quad \zeta_0^{(1)}(i, j) = 0. \qquad (6.19)$$

Remark Obviously, at $t = 0$, x_0 is independent of (A, B), making $dx_0/dB = 0$ and hence the setting $\zeta_0^{(1)} = 0$.

We show that, under the conditions of Theorem 3.4, one has

$$\lim_{t \to \infty} \Pi_t^{(1)}(i, j) = \frac{dF^T(\theta_\infty)}{dB_{ij}}, \quad i = 1, 2 \cdots n;\ j = 1, 2 \cdots m, a.s. \qquad (6.20)$$

Introducing (6.19) in (6.18) gives

$$\Pi_t^{(1)}(i, j) = P_t^{-1}\Big\{ \int_0^t \zeta_s^{(1)}(i, j)x_s^T (F^{*T}(\theta_s) - \Gamma_t)ds$$

$$+ \int_0^t x_s \zeta_s^{(1)T}(i, j)(F^T(\theta_s) - \Gamma_t)ds$$

$$+ \int_0^t \zeta_s^{(1)}(i, j)dw_s^T C^T + \int_0^t x_s x_s^T \Pi_s^{(0)} ds\Big\}$$

$$= P_t^{-1}\Big\{ \int_0^t (\zeta_s^{(1)}(i, j)x_s^T + x_s \zeta_s^{(1)T}(i, j))(F^T(\theta_s) - \Gamma_t)ds$$

$$+ \int_0^t \zeta_s^{(1)}(i, j) x_s^T (F^*(\theta_s) - F(\theta_s))^T ds + \int_0^t \zeta_s^{(1)}(i, j) dw_s^T C^T$$

$$+ \int_0^t x_s x_s^T \Big(\Pi_s^{(0)}(i, j) - \frac{dF^T(\theta_\infty)}{dB_{ij}} \Big) ds \Big\} + \frac{dF^T(\theta_\infty)}{dB_{ij}}. \qquad (6.21)$$

Taking the limit of the next-to-last term on the RHS of (6.21) gives

$$\lim_{t\to\infty} P_t^{-1} \| \int_0^t x_s x_s^T ds \Big(\Pi_s^{(0)}(i, j) - \frac{dF^T(\theta_\infty)}{dB_{ij}} \Big) ds \|$$

$$\leq \lim_{t\to\infty} P_t^{-1} \{ \| \int_0^r x_s x_s^T ds \Big(\Pi_s^{(0)}(i, j) - \frac{dF^T(\theta_\infty)}{dB_{ij}} \Big) ds \|$$

$$+ \| \int_r^t x_s x_s^T ds \| \sup_{s\geq r} \| \Pi_s^{(0)} - \frac{dF^T(\theta_\infty)}{dB_{ij}} \| \}$$

$$= \lim_{t\to\infty} P_t^{-1} \| \int_r^t x_s x_s^T ds \| \sup_{s\geq r} \| \Pi_s^{(0)} - \frac{dF^T(\theta_\infty)}{dB_{ij}} \|$$

$$\leq I \sup_{s\geq r} \| \Pi_s^{(0)} - \frac{dF^T(\theta_\infty)}{dB_{ij}} \|, \qquad (6.22)$$

which, after taking $r \to \infty$, while recalling (6.17), results in

$$\lim_{t\to\infty} P_t^{-1} \int_0^t x_s x_s^T ds \Big(\Pi_s^{(0)}(i, j) - \frac{dF^T(\theta_\infty)}{dB_{ij}} \Big) ds = 0, \ a.s. \qquad (6.23)$$

Next, to show that the first three terms on the RHS of (6.21) vanish at the limit as $t \to \infty$, we utilize (6.4): The fact that the $\{\Pi_t^{(0)}\}$ trajectory's continuity and convergence imply that $\sup_{t\geq 0} \| \Pi_t^{(0)} \| < \infty$, a.s. together with property (6.4), makes the input to the $\{\zeta_t^{(1)}\}$ (linear) ODE (6.19) uniformly bounded, a.s. resulting in

$$\sup_{t\geq 0} \| \zeta_t^{(1)}(i, j) \| < \infty, \ a.s. \qquad (6.24)$$

The proof of (6.24) rests on some of arguments leading to the corresponding statement (6.4), the difference being that here, the convolution expression of $\zeta_t^{(1)}$ consists of a standard Lebesgue integral (as opposed to a stochastic integral in the case of x_t). That convolution integral is kept uniformly bounded by virtue of the asymptotically stable limit $F(\theta_\infty)$ itself, giving rise to a transition matrix process uniformly bounded by a decaying exponential, from some a.s. finite (random) time on—see (6.46) below (see also [9, Equation (29)]). Then, given that no finite escape time exists for linear ODEs driven by bounded inputs, (6.24) is substantiated. Further details of the straightforward, yet lengthy calculations involved are omitted. (Note the dynamic matrix processes $\{F^*(\theta_t), \ t \geq 0\}$, $\{F(\theta_t), \ t \geq 0\}$ generating x_t

and $\zeta_t^{(1)}$, respectively, converge to the *same* limit, as stipulated by the convergence of θ_t to a limit in $\mathcal{I}$.)

Given (6.24), one has that $\int_0^t \|\zeta_s^{(1)}\|^2 ds = \mathcal{O}(t) = \mathcal{O}(\|P_t\|)$ (with the second equality following (6.5)). This and a martingale LLN together give

$$\lim_{t\to\infty} P_t^{-1} \int_0^t \zeta_s^{(1)} dw_s^T C^T = 0, \ a.s. \tag{6.25}$$

Consider the first term on the RHS of (6.21). To show that it vanishes at the limit, write

$$\lim \sup_{t\to\infty} P_t^{-1} \| \int_0^t \zeta_s^{(1)} x_s^T (F^T(\theta_s) - \Gamma_t) ds \|$$

$$\leq \quad \lim \sup_{t\to\infty} P_t^{-1} \| \int_0^r \zeta_s^{(1)} x_s^T (F^T(\theta_s) - \Gamma_t) ds \|$$

$$+ \lim \sup_{t\to\infty} P_t^{-1} \| \int_r^t \zeta_s^{(1)} x_s^T (F^T(\theta_s) - \Gamma_t) ds \|$$

$$= \quad \lim \sup_{t\to\infty} P_t^{-1} \| \int_r^t \zeta_s^{(1)} x_s^T (F^T(\theta_s) - \Gamma_t) ds \|$$

$$\leq \quad \lim \sup_{t\to\infty} P_t^{-1} \sqrt{\int_r^t \|\zeta_s^{(1)}\|^2 ds \int_r^t \|x_s\|^2 ds} \sup_{s\geq r} \|(F^T(\theta_s) - \Gamma_t)\|$$

$$= \quad 0, \ a.s., \tag{6.26}$$

where the last equality is obtained with $\lim \sup_{t\to\infty} P_t^{-1} \sqrt{\int_r^t \|\zeta_s^{(1)}\|^2 ds \int_r^t \|x_s\|^2 ds} = \mathcal{O}(1)$ (due to (6.5) and (6.24)), and then, by taking $r \to \infty$ while recalling (6.11). With the first term on the RHS of (6.21) diminishing to zero as $t \to \infty$, a.s., then, using similar calculations, property (6.3) leads to the vanishing of the second term on the RHS of (6.21), as $t \to \infty$, hence with (6.25), finally yielding (6.20).

6.5   Generalized Stochastic Picard Iterations

For $n \geq 2$, define the iteration

$$\Pi_t^{(n)}(i, j) = P_t^{-1} \Big[\int_0^t \zeta_s^{(n)}(i, j) dx_s^T + \int_0^t x_s d\zeta_s^{(n)T}(i, j)$$

$$- \int_0^t (\zeta_s^{(n)}(i, j) x_s^T + x_s \zeta_s^{(n)T}(i, j)) ds \Gamma_t \Big], \tag{6.27}$$

$$\dot\zeta_t^{(n)}(i, j) = F(\theta_t)\zeta_t^{(n)}(i, j) + \Pi_t^{(n-1)T}(i, j)x_t, \quad \zeta_0^{(n)}(i, j) = 0. \tag{6.28}$$

The convergence of the functions $\{\zeta_t^{(n)}, \ \Pi_t^{(n)}, \ t \geq 0, \ n \geq 0\}$ (as $n \to \infty$), to the solution of (5.24), (5.25), will be established using a generalized Picard argument which shall employ the properties of the $n = 0, 1$ cases.

By the induction argument below, we show that $\forall n \geq 2$,

$$\lim_{t\to\infty}\|\zeta_t^{(n)}(i, j) - \zeta_t^{(n-1)}(i, j)\| = 0, \ \ a.s., \tag{6.29}$$

and

$$\lim_{t\to\infty}\Pi_t^{(n)}(i, j) = \frac{d}{dB_{ij}}F^T(\theta_\infty), \ \ a.s. \tag{6.30}$$

6.6 Convergence of $\{\Pi_t^n, \zeta_t^n \ t \geq 0\}$ to Solutions of (5.10, 5.11)

Assume that (6.30) holds for all superscript indices up to $n - 1$. Let $\Delta\zeta_t^{(n)} \triangleq \zeta_t^{(n)} - \zeta_t^{(n-1)}$ (where, to simplify notation, from this point on, the indices (i, j) are omitted, unless specifically required), and note that due to (6.28) $\Delta\zeta^{(n)}$ satisfies the ODE.

$$\frac{d}{dt}\Delta\zeta_t^{(n)} = F(\theta_t)\Delta\zeta_t^{(n)} + \Delta\Pi_t^{(n-1)T}x_t, \quad \Delta\zeta_0^{(n)} = 0, \ \ n \geq 2, \tag{6.31}$$

where

$$\Delta\Pi_t^{(n)} \triangleq \Pi_t^{(n)} - \Pi_t^{(n-1)} = P_t^{-1}\left[\int_0^t \Delta\zeta_s^{(n)}dx_s^T + \int_0^t x_s d\Delta\zeta_s^{(n)T}\right.$$
$$\left. - \int_0^t (\Delta\zeta_s^{(n)}x_s^T + x_s\Delta\zeta_s^{(n)T})ds\Gamma_t\right], \ \ n \geq 2, \tag{6.32}$$

with the second equality following from (6.27). Solving (6.31) for $\Delta\zeta^{(n)}$ results in

$$\Delta\zeta_t^{(n)} = \int_0^t \Phi(t, s)\Delta\Pi_s^{(n-1)T}x_s ds = \Phi(t, T)\Delta\zeta_T^{(n)} + \int_T^t \Phi(t, s)\Delta\Pi_s^{(n-1)T}x_s ds, \tag{6.33}$$

where $\Phi(t, s)$ is the state transition matrix from s to t, generated by $\{F(\theta_r), \ s \leq r \leq t\}$.

To show that $\Delta\zeta_t^{(n)} \to 0$, $as \ t \to \infty$, $a.s.$, we replicate the arguments leading to the asymptotic vanishing of the solution to the linear ODE with a diminishing input (5.52): First, observe that the input to the ODE (6.31) is diminishing, namely, $\Delta \Pi_t^{(n-1)T} x_t \to 0$, as $t \to \infty$. This is due to the induction argument above that (6.30) holds for all indices up to $n - 1$, together with the boundedness of $\{x_t\}$ (6.4). Then, since $F(\theta_t) \to F(\theta_\infty)$ as $t \to \infty$, a.s. with the limit $F(\theta_\infty)$ Hurwitz, clearly, there exists a $\tau_F = \tau_F(\omega) < \infty$, such that $\lambda_{max}(F(\theta_t)) \leq \frac{1}{2}\lambda_{max}(F(\theta_\infty)) < 0$, for all $t \geq \tau_F$, a.s. With that, one observes that the perturbed linear ODE (6.31) satisfies the conditions of Khalil [22, Lemma 5.4], thereby making (6.31) locally input-to-state stable [22, Definition 5.2], for all initial times $t_0 \geq \tau_F$, a.s. (We note again that here t_0 refers to the notation given in [22, Definition 5.2], not to be confused with the initial time t_0 stated in this work.) Then, one may take $t = 2t_0$ in [22, Definition 5.2] to conclude that

$$\Delta \zeta_t^{(n)} \to 0, \ as \ t \to \infty, \ a.s. \tag{6.34}$$

Consider $\Delta \Pi_t^{(n)}$ given by (6.32). Using the fact that, with $\Delta \zeta_0^{(n)} = 0$,

$$\| \int_0^t \Delta \zeta_s^{(n)} dx_s^T + \int_0^t x_s d\Delta \zeta_s^{(n)T} \| = \|\Delta \zeta_t^{(n)} x_t^T\|,$$

then, with $P_t = \int_0^t x_s x_s^T ds$, splitting the third integral on the RHS of (6.32) into two integrals (one over $[0, T]$ and the second over $[T, t]$) gives

$$\|\Delta \Pi_t^{(n)}\| \leq \frac{2}{\lambda_{min}(P_t)} \Big[\|x_t\| + \sqrt{(t - T)\mathrm{Tr}(P_t - P_T)}\|\Gamma_t\|\Big] \sup_{T \leq s \leq t} \|\Delta \zeta_s^{(n)}\|$$
$$+ \frac{2}{\lambda_{min}(P_t)} \sqrt{T\mathrm{Tr}P_T}\|\Gamma_t\| \sup_{0 \leq s \leq T} \|\Delta \zeta_s^{(n)}\|. \tag{6.35}$$

Using (6.34), fix $\varepsilon > 0$ and choose a $T < \infty$ such that $\sup_{s \geq T} \|\Delta \zeta_s^{(n)}\| < \varepsilon$. With $\lambda_{min}(P_t) = \mathcal{O}(\|P_t\|) = \mathcal{O}(t)$, it follows from (6.32) that

$$\lim \sup_{t \to \infty} \|\Delta \Pi_t^{(n)}\| = \mathcal{O}(\varepsilon), \ a.s.$$

Since this holds for any $\varepsilon > 0$, the conclusion that (6.30) holds for n follows.

For $T = 0$, (6.35) is written in the form,

$$\|\Delta \Pi_t^{(n)}\| \leq \frac{2}{\lambda_{min}(P_t)} \Big(\|x_t\| + \sqrt{t\mathrm{Tr}P_t}\|\Gamma_t\|\Big) \sup_{0 \leq s \leq t} \|\Delta \zeta_s^{(n)}\|, \tag{6.36}$$

where (6.4), (6.7), and (6.11) (itself making $\sup_{t \geq t_0} \|\Gamma_t\| < \infty$, a.s.) together imply that

$$a = a(\omega) \triangleq \sup_{t \geq t_0} \frac{2}{\lambda_{min}(P_t)} (\|x_t\| + \sqrt{t\mathrm{Tr}P_t}\|\Gamma_t\|) < \infty, \ a.s. \tag{6.37}$$

(for a $t_0 > 0$ such that $\lambda_{min}(P_{t_0}) \geq \varepsilon$, some $\varepsilon > 0$). Then, for any $t \geq t_0$,

$$\sup_{0 \leq s \leq t} \|\Delta\zeta_s^{(n+1)}\| \leq a \int_0^t \|\Phi(t,s)\| \sup_{0 \leq r \leq s} \|\Delta\zeta_r^{(n)}\| ds \leq \rho \int_0^t \sup_{0 \leq r \leq s} \|\Delta\zeta_r^{(n)}\| ds,$$

$$(6.38)$$

where $\rho \overset{\Delta}{=} a\kappa \sup_{t \geq t_0} \|x_t\|$, with $\kappa = \sup_{t \geq s \geq 0} \|\Phi(t,s)\|$.

Lemma 6.4 *Given that $F(\theta_t) \to F(\theta_\infty)$, Hurwitz, as $t \to \infty$, a.s., one has that*

$$\kappa = \sup_{t \geq s \geq 0} \|\Phi(t,s)\| < \infty, \quad a.s. \tag{6.39}$$

Proof Consider the autonomous version of (6.31),

$$\dot{z}_t = F(\theta_t)z_t. \tag{6.40}$$

Fix a positive-definite matrix Q and let $v(x) = x^T P x$, with P the unique solution of

$$F^T(\theta_\infty)P + PF(\theta_\infty) + Q = 0. \tag{6.41}$$

As $F(\theta_\infty)$ is Hurwitz, it is well known that $P > 0$. Then, with

$$D_t = F(\theta_t) - F(\theta_\infty) \to 0, \quad a.s., \ as \ t \to \infty, \tag{6.42}$$

one has that

$$\begin{aligned}
\dot{v}(z) &= z^T(F(\theta_t)^T P + PF(\theta_t))z + z^T(D_t^T P + PD_t)z \\
&= -z^T Q z + z^T \Delta(t)z, \tag{6.43}
\end{aligned}$$

where by (6.42) defining D_t,

$$\Delta(t) = PD_t + D_t^T P \to 0, \quad a.s., \ as \ t \to \infty. \tag{6.44}$$

Choose a $\delta > 0$ such that $Q/2 - \delta I > 0$ and set $T = T(\omega) = \sup\{t > 0 : Q/2 - \Delta(t) < \delta I\}$. (Obviously, $T \in \mathcal{F}_\infty^x$ is not a stopping time.) Then, (6.44) together with the choice of δ implies that $T < \infty$, a.s., and

$$\dot{v}(z) = -z^T(Q - \Delta(t))z \leq -\|z\|^2 \delta, \ \forall t \geq T, \ a.s. \tag{6.45}$$

Property (6.45) implies that the equilibrium point $z = 0$ of (6.40) is (globally) $[T, \infty)$-uniformly asymptotically stable, a.s. Hence, by Theorem 4.2 of Khalil [22], the state transition matrix process $\{\Phi(t, s),\ t \geq s \geq T\}$ generated by $\{F(\theta_r),\ r \geq T\}$, satisfies

$$\|\Phi(t, s)\| \leq k_1 \exp^{-\eta(t-s)}, \quad \forall t \geq s \geq T, \tag{6.46}$$

for some (possibly random) constants $k_1, \eta > 0$. Since, obviously,

$$\sup_{0 \leq s \leq t \leq T} \|\Phi(t, s)\| = k_0 < \infty, \ a.s., \tag{6.47}$$

it follows that (6.39) holds with $\kappa = k_0 k_1$.   □

Then, given (6.38) together with (6.39),

$$\sup_{0 \leq s \leq t} \|\Delta \zeta_s^{(n+1)}\| \leq \rho \int_0^t \sup_{0 \leq r \leq s} \|\Delta \zeta_r^{(n)}\| ds \leq \rho^2 \int_0^t \int_0^s \sup_{0 \leq r \leq u} \|\Delta \zeta_r^{(n-1)}\| du ds$$

$$\leq \cdots \leq \rho^n \int_0^t \cdots \int_0^q \sup_{0 \leq r \leq v} \|\Delta \zeta_r^{(1)}\| dv \cdots ds \leq \frac{(\rho t)^n}{n!} \sup_{0 \leq s \leq t} \|\Delta \zeta_s^{(1)}\|. \tag{6.48}$$

And, with (6.36), (6.37), and (6.48), one also has

$$\sup_{0 \leq s \leq t} \|\Delta \Pi_s^{(n+1)}\| \leq a \sup_{0 \leq s \leq r} \|\Delta \zeta_s^{(n+1)}\| \leq a \frac{(\rho t)^n}{n!} \sup_{0 \leq s \leq t} \|\Delta \zeta_s^{(1)}\|. \tag{6.49}$$

With (6.48), (6.49), and the summability of $\{(\rho t)^n / n!\}_{n \geq 0}$, it is now obvious that the series $\{\zeta_s^{(n)},\ \Pi_s^{(n)},\ s \in [0, t],\ n \geq 0\}$ is Cauchy on $\mathcal{C}[0, t]$, equipped with the sup-norm (and thus complete), namely,

$$\lim_{n, m \to \infty} \sup_{s \in [0, t]} \|(\zeta_s^{(n)}, \Pi_s^{(n)}) - (\zeta_s^{(m)}, \Pi_s^{(m)})\| = 0, \ a.s., \tag{6.50}$$

and hence,

$$\{\zeta_s^{(n)}, \Pi_s^{(n)},\ s \in [0, t]\}$$
$$\to \{\zeta_s, \Pi_s,\ s \in [0, t]\}, \quad \text{uniformly on } [0, t], \text{ as } n \to \infty, \ a.s. \tag{6.51}$$

To extend (6.51) to $[0, \infty)$, we consider intervals $[T, t]$ with T being sufficiently large. Here we take T as set below (6.44).

Solving (6.31) for $\Delta \zeta^{(n+1)}$ results in

$$\Delta \zeta_t^{(n+1)} = \Phi(t, T) \Delta \zeta_T^{(n+1)} + \int_T^t \Phi(t, s) \Delta \Pi_s^{(n)T} x_s ds, \tag{6.52}$$

where, as above, $\Phi(t, s)$ is the state transition matrix from s to t, generated by $\{F(\theta_r), \; s \leq r \leq t\}$. Introducing (6.35) in equation (6.52), while utilizing (6.46), results in

$$\|\Delta\zeta_t^{(n+1)}\| \leq e^{-\eta(t-T)}\rho T \sup_{0 \leq s \leq T} \|\Delta\zeta_s^{(n)}\| +$$

$$+ e^{-\eta t} \int_T^t e^{\eta s}\Big[\tilde{c}(s, T) \sup_{0 \leq r \leq T} \|\Delta\zeta_s^{(n)}\| + \tilde{a}(s, T) \sup_{T \leq r \leq s} \|\Delta\zeta_r^{(n)}\|\Big]ds,$$

$$(6.53)$$

where the following definitions are being used:

$$\tilde{c}(s, T) \triangleq \frac{2\kappa\|x_s\|}{\lambda_{min}(P_s)}\sqrt{T\mathrm{Tr}P_T}\|\Gamma_s\| = \mathcal{O}(T/s),$$

$$\tilde{a}(s, T) \triangleq \frac{2\kappa\|x_s\|}{\lambda_{min}(P_s)}[\|x_s\| + \sqrt{(s-T)\mathrm{Tr}(P_s - P_T)}\|\Gamma_s\|] = \mathcal{O}((s-T)/s),$$

with the $\mathcal{O}(\cdot)$ equalities due to the fact that, by (6.5), $\lambda_{min}(P_t) = \mathcal{O}(\|P_t\|) = \mathcal{O}(t)$. Note that both $\tilde{a}(s, T)$ and $\tilde{c}(s, T)$ may be upper bounded by

$$\tilde{c}(s, T) \leq c\frac{T}{s}, \quad c = c(\omega) = \sup_{s>T>0} \frac{s}{T}\tilde{c}(s, T) < \infty, \; a.s.$$

$$\tilde{a}(s, T) \leq b\frac{s-T}{s}, \quad b = b(\omega) = \sup_{s>T>0} \frac{s}{s-T}\tilde{a}(s, T) < \infty, \; a.s.$$

With these bounds, (6.53) may be replaced by

$$\|\Delta\zeta_t^{(n+1)}\| \leq e^{-\eta t}\Big\{e^{\eta T}\rho T \sup_{0 \leq s \leq T} \|\Delta\zeta_s^{(n)}\| +$$

$$+ \int_T^t \frac{e^{\eta s}}{s}\Big[cT \sup_{0 \leq r \leq T} \|\Delta\zeta_r^{(n)}\| + b(s-T) \sup_{T \leq r \leq s} \|\Delta\zeta_r^{(n)}\|\Big]ds\Big\} \leq$$

$$\leq 3\max\Big\{e^{-\eta(t-T)}\rho T \sup_{0 \leq s \leq T} \|\Delta\zeta_s^{(n)}\|,$$

$$e^{-\eta t}\int_T^t \frac{e^{\eta s}}{s}b(s-T) \sup_{T \leq r \leq s} \|\Delta\zeta_r^{(n)}\|ds,$$

$$e^{-\eta t}\int_T^t \frac{e^{\eta s}}{s}cT \sup_{0 \leq r \leq T} \|\Delta\zeta_r^{(n)}\|ds\Big\}. \qquad (6.54)$$

Taking the sup over $[T, t]$, one obtains

$$\sup_{T \leq s \leq t} \|\Delta\zeta_s^{(n+1)}\| \leq 3 \max \Big\{\rho T \sup_{0 \leq s \leq T} \|\Delta\zeta_s^{(n)}\|, \frac{b}{\eta} \sup_{T \leq r \leq t} \|\Delta\zeta_r^{(n)}\|,$$

$$\sup_{T \leq u \leq t} e^{-\eta u} \int_T^u \frac{e^{\eta s}}{s} ds \, cT \sup_{0 \leq r \leq T} \|\Delta\zeta_r^{(n)}\|\Big\}, \qquad (6.55)$$

and further, since $\sup_{t \geq T} cT e^{-\eta t} \int_T^t (e^{\eta s}/s) ds \leq c/\eta$, it follows that for any $T \geq c/\rho\eta$,

$$\sup_{T \leq s \leq t} \|\Delta\zeta_s^{(n+1)}\| \leq 3 \max \Big\{\rho T \sup_{0 \leq s \leq T} \|\Delta\zeta_s^{(n)}\|, \frac{b}{\eta} \sup_{T \leq r \leq t} \|\Delta\zeta_r^{(n)}\|\Big\}. \qquad (6.56)$$

It is therefore the case that, by taking $t \to \infty$, one has

$$\sup_{T \leq s < \infty} \|\Delta\zeta_s^{(n+1)}\| \leq 3 \max \Big\{\rho T \sup_{0 \leq s \leq T} \|\Delta\zeta_s^{(n)}\|, \frac{b}{\eta} \sup_{T \leq s < \infty} \|\Delta\zeta_s^{(n)}\|\Big\}. \qquad (6.57)$$

Let

$$T_n = T_n(\omega) \stackrel{\Delta}{=} \inf\{T \geq 0 : \rho T \sup_{0 \leq s \leq T} \|\Delta\zeta_s^{(n)}\| \geq \frac{b}{\eta} \sup_{T \leq s < \infty} \|\Delta\zeta_s^{(n)}\|\}. \qquad (6.58)$$

Then, obviously, due to (6.29), one has that for each n, $T_n < \infty$, a.s. We now show that $\sup_{n \geq 1} T_n < \infty$, a.s. Suppose not. Then, since $P(T_n < \infty, \ \forall n \geq 1) = 1$, it follows that $\limsup_{n \to \infty} T_n = \infty$, a.e. on an event $\widetilde{\Omega}$, with $P(\widetilde{\Omega}) > 0$. Next, note that by definition,

$$T_n = \frac{b}{\rho\eta} \sup_{s \in [T_n, \infty)} \|\Delta\zeta_s^{(n)}\| / \sup_{s \in [0, T_n]} \|\Delta\zeta_s^{(n)}\|.$$

Extract an increasing subsequence $\{T_{n_k}\}$ such that $T_{n_k} \uparrow \infty$ a.e. on $\widetilde{\Omega}$, as $k \to \infty$. Then, a.e. on $\widetilde{\Omega}$, one has

$$\infty = \lim_{k \to \infty} T_{n_k} = \lim_{k \to \infty} \frac{b}{\rho\eta} \sup_{s \in [T_{n_k}, \infty)} \|\Delta\zeta_s^{(n_k)}\| / \sup_{s \in [0, T_{n_k}]} \|\Delta\zeta_s^{(n_k)}\|$$

$$\leq \frac{b}{\rho\eta} \lim_{k \to \infty} \bigvee_{m=1}^{n_k} \Big(\sup_{s \in [T_{n_k}, \infty)} \|\Delta\zeta_s^{(m)}\| / \sup_{s \in [0, T_{n_k}]} \|\Delta\zeta_s^{(m)}\| \Big) < \frac{b}{\rho\eta} < \infty, \qquad (6.59)$$

where the last inequality is due to the fact that $T_{n_k} \uparrow \infty$. With this contradiction, one finally has that $P(\sup_{n \geq 1} T_n < \infty) = 1$, and hence, with the aid of (6.48), one may conclude that for all $T \geq \sup_{n \geq 1} T_n$,

$$\sup_{T \leq s < \infty} \|\Delta \zeta_s^{(n+1)}\| \leq 3\rho T \sup_{0 \leq s \leq T} \|\Delta \zeta_s^{(n)}\| \leq 3\rho T \frac{(\rho T)^{(n-1)}}{(n-1)!} \sup_{0 \leq s \leq T} \|\Delta \zeta_s^{(1)}\|.$$

$$(6.60)$$

Following (6.36), (6.37), (6.48), and (6.60), one has

$$\sup_{T \leq s < \infty} \|\Delta \Pi_s^{(n+1)}\|$$

$$\leq a \sup_{0 \leq s < \infty} \|\Delta \zeta_s^{(n+1)}\|$$

$$= a \max \Big\{ \sup_{0 \leq s \leq T} \|\Delta \zeta_s^{(n+1)}\|, \ \sup_{T \leq s < \infty} \|\Delta \zeta_s^{(n+1)}\| \Big\}$$

$$\leq a \max \Big\{ \frac{(\rho T)^n}{n!} \sup_{0 \leq s \leq T} \|\Delta \zeta_s^{(1)}\|, \ 3\rho T \frac{(\rho T)^{(n-1)}}{(n-1)!} \sup_{0 \leq s \leq T} \|\Delta \zeta_s^{(1)}\| \Big\}$$

$$= 3a\rho T \frac{(\rho T)^{(n-1)}}{(n-1)!} \sup_{0 \leq s \leq T} \|\Delta \zeta_s^{(1)}\|.$$

$$(6.61)$$

With (6.60), (6.61), utilizing, as above, the fact that $\{(\rho T)^{(n-1)}/(n-1)!, \ n \geq 1\}$ is summable, one concludes that $\{\zeta_s^{(n)}, \ \Pi_s^{(n)}, s \in [T, \infty), \ n \geq 0\}$ is Cauchy in $\mathcal{C}[T, \infty)$ (under the sup-norm, thus complete) and hence uniformly converges to $\{\zeta_s, \ \Pi_s, \ s \in [T, \infty)\}$, a.s. Combining this property together with (6.51) yields

$$\{\zeta_t^{(n)}, \Pi_t^{(n)}, \ t \geq 0\} \to \{\zeta_t, \Pi_t, \ t \geq 0\}, \quad \text{uniformly on } [0, \infty) \text{ as } n \to \infty, \ a.s.,$$

$$(6.62)$$

where Π_t and ζ_t are defined in (5.10) and (5.11) (respectively).

The Consistency of $\{\Pi_t, \ t \geq 0\}$

Finally,

$$\|\Pi_t(i, j) - \frac{d}{dB_{ij}} F^T(\theta_\infty)\|$$

$$\leq \|\Pi_t^{(n)}(i, j) - \frac{d}{dB_{ij}} F^T(\theta_\infty)\| + \|\Pi_t(i, j) - \Pi_t^{(n)}(i, j)\|.$$

$$(6.63)$$

Fix some $\varepsilon > 0$ and take $n < \infty$ such that, by (6.62),

$$\sup_{t \geq 0} \|\Pi_t(i, j) - \Pi_t^{(n)}(i, j)\| < \varepsilon, \quad a.s. \tag{6.64}$$

Then, in the first term on the RHS of (6.63), take $t \to \infty$ to yield (by (6.30)),

$$\limsup_{t \to \infty} \|\Pi_t^{(n)}(i, j) - \frac{d}{dB_{ij}} F^T(\theta_\infty)\| = 0, \quad a.s., \tag{6.65}$$

and hence,

$$\limsup_{t \to \infty} \|\Pi_t(i, j) - \frac{d}{dB_{ij}} F^T(\theta_\infty)\| < \varepsilon, \quad a.s. \tag{6.66}$$

As this holds for any $\varepsilon > 0$, the proof of (5.12) is now complete. $\square$

References

1. Åström, K.J. and Wittenmark, B., On Self-tuning regulators. *Automatica*, Vol 9, (1973) 185–199
2. Becker, A., Kumar, P. R. and Wei, C. Z. Adaptive control with the stochastic approximation algorithm: geometry and convergence, *IEEE Trans. Automatic Control,* Vol AC-30(4), (1985) 330–338.
3. Borkar, V.S., Self-Tuning Control of Diffusions without the Identifiability Condition. *J. Opt. Theory & Appl*, Vol. 68 (1991) 117–138.
4. Caines, P.E. *Linear Stochastic Systems* (John Wiley, 1988; republished in SIAM Classics 2018).
5. Caines, P.E., Continuous-Time Stochastic Adaptive Control: Non-explosion, ∈-Consistency and Stability. *Systems & Control Letters* 19 (1992) 169–176.
6. Caines, P.E. and Lafortune S., "Adaptive Control with Recursive Identification for Stochastic Linear Systems". *IEEE Trans. on Automatic Control.* Vol.AC-29, No. 4, (April 1984), 312–321.
7. Caines, P.E. and Levanony, D., Stochastic Linear Quadratic Adaptive Control: A Conceptual Scheme. *Proceedings of the 44th IEEE Conference on Decision and Control* (2005)
8. Caines, P.E. and Levanony, D., Stochastic ∈-Optimal LQ Adaptation: An Alternating Controls Policy. *SIAM J. Control and Optimization* Vol. 57, No. 2 (2019) 1094–1126.
9. Caines, P.E. and Levanony, D., On bounded solutions of linear SDEs driven by convergent dynamic matrix processes with Hurwitz limits. *Stochastics* Vol. 93, No. 6 (2021) 857–867.
10. Campi, M.C. and Kumar, P.R., Adaptive Linear Quadratic Adaptive Control: The Cost–Biased Approach Revisited, *SIAM J. Control and Optimization*, Vol. 36 (1998) 1890–1907.
11. Chen, H.F. and Guo, L., *Identification and Stochastic Adaptive Control*, Birkhöuser, (1991).
12. Clarke, D. and Gawthrop, P. Self-tuning control. *Proc. IEE*, Vol 122(a), (1975) 929–934.
13. Duncan, T.E. and Pasik-Duncan, B., Adaptive Control of Continuous–Time Linear Stochastic Systems, *Mathematics of Control, Signals and Systems* 3 (1990) 45–60.
14. Duncan, T.E. and Pasik-Duncan, B., A Parameter Estimate Associated with the Adaptive Control of Stochastic Systems. In: *Analysis and Optimization of Systems*, L.N. Control & Inf. Sc. 83 (Springer, 1986) 508–514.
15. Duncan, T.E. and Pasik-Duncan, B., Some Methods for the Adaptive Control of Continuous Time Linear Stochastic Systems. In: *Topics in Stochastic Systems: Modelling, Estimation and Adaptive Control*, Gerencsér, L. and Caines, P.E., Eds. L.N. Control & Info. Sc. 161 (Springer, 1991).
16. Duncan, T.E., Guo, L. and Pasik-Duncan, B., Adaptive Continuous–Time Linear Quadratic Gaussian Control. *IEEE Trans. Automatic Control* 44 (1999) 1653–1662.

© The Author(s), under exclusive license to Springer Nature Switzerland AG 2024
D. Levanony, P. E. Caines, *Stochastic Lagrangian Adaptation*, SpringerBriefs in Mathematics, https://doi.org/10.1007/978-3-031-73758-9

17. Faradonbeh, M.K.S. and Faradonbeh, M.S.S., Reinforcement Learning Policies in Continuous-Time Linear Systems, *arXiv:2109.07630v3* (2023).
18. Goodwin, G. C. and Sin, K. S. *Adaptive Filtering, Prediction and Control* Prentice-Hall, Englewood Cliffs, N.J. (1984)
19. Goodwin, G.C., P.R. Ramadge and P.E. Caines, "Discrete Time Stochastic Adaptive Control". *SIAM J. Control and Optimization*, Vol. 19, No. 6, November (1981), pp. 829–853. Erratum: Vol. 20, No. 6, p. 893, November (1982).
20. J. Jacod and A.N. Shiryayev, *Limit Theorems for Stochastic Processes*. (Springer, New York, 1987).
21. Kallianpur,G., *Stochastic Filtering Theory*. (Springer, New York, 1980).
22. Khalil, H.K., *Nonlinear Systems* (Macmillan, 1992).
23. Kumar, P.R. and Becker, A., A New Family of Optimal Adaptive Controllers for Markov Chains. *IEEE Trans. Automatic Control*, Vol. AC-27 (1982) 137–146.
24. Kumar, P.R. and Lin, W., Optimal Adaptive Controllers for Unknown Markov Chains. *IEEE Trans. Automatic Control*, Vol. AC-27 (1982) 765–774.
25. Kumar, P.R., Optimal Adaptive Control of Linear-Quadratic-Gaussian Systems, *SIAM J. Control and Optimization*, Vol. 21 (1983) 163–178.
26. Kumar, P.R., Convergence of Adaptive Control Schemes Using Least Squares Estimates. *Proceedings of the 28th Conference on Decision and Control* (1989), 727–731.
27. Kunita, H., *Stochastic Flows and Stochastic Differential Equations*, (Cambridge University Press, 1990).
28. Lai, T.L. and Wei, C.Z., Least Squares Estimates in Stochastic Regression Models with Application to Identification and Control of Dynamic Systems, *Ann. Stat.* 10 (1982) 154–166.
29. Lai, T.L. and Wei, C.Z., Asymptotically Efficient Self-Tuning Regulators. *SIAM J. Control Opt.*, Vol. 25 (1987) 466–481.
30. Levanony, D., Shwartz, A. and Zeitouni, O., Recursive Identification in Continuous-Time Stochastic Processes. *Stochastic Proc. & Appl.*, Vol. 49 (1994) 245–275.
31. Levanony D. and Caines, P.E., On Persistent Excitation for Linear Systems with Stochastic Coefficients. *SIAM J. Control and Optimization* Vol. 40 (2001) 882–897.
32. Levanony D. and Caines, P.E., Stochastic Lagrangian Adaptive LQG Control. In *Lecture Notes in Control and Information Sciences*, Pasik-Duncan, B. (Ed.), Vol. 280, (Springer, 2002) 283–300.
33. Levanony D., On the Consistent Filtering of Convergent Semimartingales. *Stochastic Processes and their Applications*, Vol. 129, Issue 1 (2019) 323–335.
34. Polderman, J.W., A Note on the Structure of Two Subsets of the Parameter Space in Adaptive Control Problems. *Systems & Control Letters* 7 (1986) 25–34.
35. Polderman, J.W., On the Necessity of Identifying the True Parameter in Adaptive LQ Control. *Systems & Control Letters* 8 (1986) 87–91.
36. Polderman, J.W., Adaptive LQ Control: Conflict Between Identification and Control. *Linear Algebra & Applications* (1989) 219–244.
37. Prandini, M. and Campi, M.C., Adaptive LQG Control of Input–Output Systems: A Cost–Biased Approach. *SIAM J. Control and Optimization*, Vol. 39 (2001) 1499–1519.
38. Revuz, D. and Yor, M., *Continuous Martingales and Brownian Motion* (Springer, 1991).
39. Wang, F. and Janson, L., Exact Asymptotics for Linear Quadratic Adaptive Control. *Journal of Machine learning Research* 22 (2021) 1–112.

17. Faradonbeh, M.K.S. and Faradonbeh, M.S.S., Reinforcement Learning Policies in Continuous-Time Linear Systems, *arXiv:2109.07630v3* (2023).

18. Goodwin, G. C. and Sin, K. S. *Adaptive Filtering, Prediction and Control* Prentice-Hall, Englewood Cliffs, N.J. (1984)

19. Goodwin, G.C., P.R. Ramadge and P.E. Caines, "Discrete Time Stochastic Adaptive Control". *SIAM J. Control and Optimization*, Vol. 19, No. 6, November (1981), pp. 829–853. Erratum: Vol. 20, No. 6, p. 893, November (1982).

20. J. Jacod and A.N. Shiryayev, *Limit Theorems for Stochastic Processes*. (Springer, New York, 1987).

21. Kallianpur,G., *Stochastic Filtering Theory*. (Springer, New York, 1980).

22. Khalil, H.K., *Nonlinear Systems* (Macmillan, 1992).

23. Kumar, P.R. and Becker, A., A New Family of Optimal Adaptive Controllers for Markov Chains. *IEEE Trans. Automatic Control*, Vol. AC-27 (1982) 137–146.

24. Kumar, P.R. and Lin, W., Optimal Adaptive Controllers for Unknown Markov Chains. *IEEE Trans. Automatic Control*, Vol. AC-27 (1982) 765–774.

25. Kumar, P.R., Optimal Adaptive Control of Linear-Quadratic-Gaussian Systems, *SIAM J. Control and Optimization*, Vol. 21 (1983) 163–178.

26. Kumar, P.R., Convergence of Adaptive Control Schemes Using Least Squares Estimates. *Proceedings of the 28th Conference on Decision and Control* (1989), 727–731.

27. Kunita, H., *Stochastic Flows and Stochastic Differential Equations*, (Cambridge University Press, 1990).

28. Lai, T.L. and Wei, C.Z., Least Squares Estimates in Stochastic Regression Models with Application to Identification and Control of Dynamic Systems, *Ann. Stat.* 10 (1982) 154–166.

29. Lai, T.L. and Wei, C.Z., Asymptotically Efficient Self-Tuning Regulators. *SIAM J. Control Opt.*, Vol. 25 (1987) 466–481.

30. Levanony, D., Shwartz, A. and Zeitouni, O., Recursive Identification in Continuous-Time Stochastic Processes. *Stochastic Proc. & Appl.*, Vol. 49 (1994) 245–275.

31. Levanony D. and Caines, P.E., On Persistent Excitation for Linear Systems with Stochastic Coefficients. *SIAM J. Control and Optimization* Vol. 40 (2001) 882–897.

32. Levanony D. and Caines, P.E., Stochastic Lagrangian Adaptive LQG Control. In *Lecture Notes in Control and Information Sciences*, Pasik-Duncan, B. (Ed.), Vol. 280, (Springer, 2002) 283–300.

33. Levanony D., On the Consistent Filtering of Convergent Semimartingales. *Stochastic Processes and their Applications*, Vol. 129, Issue 1 (2019) 323–335.

34. Polderman, J.W., A Note on the Structure of Two Subsets of the Parameter Space in Adaptive Control Problems. *Systems & Control Letters* 7 (1986) 25–34.

35. Polderman, J.W., On the Necessity of Identifying the True Parameter in Adaptive LQ Control. *Systems & Control Letters* 8 (1986) 87–91.

36. Polderman, J.W., Adaptive LQ Control: Conflict Between Identification and Control. *Linear Algebra & Applications* (1989) 219–244.

37. Prandini, M. and Campi, M.C., Adaptive LQG Control of Input–Output Systems: A Cost–Biased Approach. *SIAM J. Control and Optimization*, Vol. 39 (2001) 1499–1519.

38. Revuz, D. and Yor, M., *Continuous Martingales and Brownian Motion* (Springer, 1991).

39. Wang, F. and Janson, L., Exact Asymptotics for Linear Quadratic Adaptive Control. *Journal of Machine learning Research* 22 (2021) 1–112.

References

1. Åström, K.J. and Wittenmark, B., On Self-tuning regulators. *Automatica*, Vol 9, (1973) 185–199
2. Becker, A., Kumar, P. R. and Wei, C. Z. Adaptive control with the stochastic approximation algorithm: geometry and convergence, *IEEE Trans. Automatic Control*, Vol AC-30(4), (1985) 330–338.
3. Borkar, V.S., Self-Tuning Control of Diffusions without the Identifiability Condition. *J. Opt. Theory & Appl*, Vol. 68 (1991) 117–138.
4. Caines, P.E. *Linear Stochastic Systems* (John Wiley, 1988; republished in SIAM Classics 2018).
5. Caines, P.E., Continuous-Time Stochastic Adaptive Control: Non-explosion, ϵ-Consistency and Stability. *Systems & Control Letters* 19 (1992) 169–176.
6. Caines, P.E. and Lafortune S., "Adaptive Control with Recursive Identification for Stochastic Linear Systems". *IEEE Trans. on Automatic Control*. Vol.AC-29, No. 4, (April 1984), 312–321.
7. Caines, P.E. and Levanony, D., Stochastic Linear Quadratic Adaptive Control: A Conceptual Scheme. *Proceedings of the 44th IEEE Conference on Decision and Control* (2005)
8. Caines, P.E. and Levanony, D., Stochastic ϵ-Optimal LQ Adaptation: An Alternating Controls Policy. *SIAM J. Control and Optimization* Vol. 57, No. 2 (2019) 1094–1126.
9. Caines, P.E. and Levanony, D., On bounded solutions of linear SDEs driven by convergent dynamic matrix processes with Hurwitz limits. *Stochastics* Vol. 93, No. 6 (2021) 857–867.
10. Campi, M.C. and Kumar, P.R., Adaptive Linear Quadratic Adaptive Control: The Cost–Biased Approach Revisited, *SIAM J. Control and Optimization*, Vol. 36 (1998) 1890–1907.
11. Chen, H.F. and Guo, L., *Identification and Stochastic Adaptive Control*, Birkhäuser, (1991).
12. Clarke, D. and Gawthrop, P. Self-tuning control. *Proc. IEE*, Vol 122(a), (1975) 929–934.
13. Duncan, T.E. and Pasik-Duncan, B., Adaptive Control of Continuous–Time Linear Stochastic Systems, *Mathematics of Control, Signals and Systems* 3 (1990) 45–60.
14. Duncan, T.E. and Pasik-Duncan, B., A Parameter Estimate Associated with the Adaptive Control of Stochastic Systems. In: *Analysis and Optimization of Systems*, L.N. Control & Inf. Sc. 83 (Springer, 1986) 508–514.
15. Duncan, T.E. and Pasik-Duncan, B., Some Methods for the Adaptive Control of Continuous Time Linear Stochastic Systems. In: *Topics in Stochastic Systems: Modelling, Estimation and Adaptive Control*, Gerencsér, L. and Caines, P.E., Eds. L.N. Control & Info. Sc. 161 (Springer, 1991).
16. Duncan, T.E., Guo, L. and Pasik-Duncan, B., Adaptive Continuous–Time Linear Quadratic Gaussian Control. *IEEE Trans. Automatic Control* 44 (1999) 1653–1662.

© The Author(s), under exclusive license to Springer Nature Switzerland AG 2024
D. Levanony, P. E. Caines, *Stochastic Lagrangian Adaptation*, SpringerBriefs in Mathematics, https://doi.org/10.1007/978-3-031-73758-9

Index

A

Adaptive feedback law, 33
Adaptive stabilization, 3
Adjoint variable, 35, 53
Algebraic Riccati equation (AER), 8
Almost everywhere (a.e.), 11, 46–48, 50, 69
A localization argument, 19–20
Alternating gain matrices, 3
AML closed-loop consistency, 23–24
A nonlinear filtering problem, 16–17
Asymptotically linearly dependent, 4
Asymptotically optimal stochastic adaptive
    control, 1–2
Asymptotically stable matrix, 29–31, 66
Asymptotic long range average optimality, 2
Asymptotic maximum likelihood (AML), 3, 5,
    11–26, 35, 45, 46
Autonomous version, 66
Autoregressive moving average (ARMAX), 1
Auxiliary process, 15
Auxiliary RLS estimate, 22

B

Bayesian embedding (BE), 11, 14, 45, 46, 49
Biased estimates, 2, 3, 43
Block diagonal matrix, 24
Bounded, 2, 4, 14, 15, 17, 23, 43–46, 48, 49,
    62, 65, 68
Bounded coefficients, 48
Bounded trajectory, 46
Bounded variation (BV), 17
Brownian motion, 7, 17, 41

C

Cauchy sequence, 57
Causal, 14
Certainty equivalence, 1–5, 8, 27, 36, 37, 43,
    44, 49
Closed loop dynamics, 3, 11, 15, 16, 27, 36,
    55
Complex conjugate, 30
Convergence, 1, 4, 13–16, 18–19, 22–23, 27,
    45, 49–50, 55, 60, 62–70
Convolution integral, 62
Cross variation, 18, 43
Cumulative quadratic cost, 54
C^ω-manifold, 38

D

Diffusion, 2, 42
Diminishing excitation signal, 4
Dither injection, 2

E

Equivalence class, 2
Event, 47, 50, 69
Exponential decay, 41, 48, 51, 54

F

Feedback gain, 15
Finite escape time, 17, 48
First-order approximation, 61–63

© The Editor(s) (if applicable) and The Author(s), under exclusive license to
Springer Nature Switzerland AG 2024
D. Levanony, P. E. Caines, *Stochastic Lagrangian Adaptation*, SpringerBriefs in
Mathematics, https://doi.org/10.1007/978-3-031-73758-9

Full derivative, 37, 39
Full rank noise, 8

G
Gaussian, 11, 14, 45
Generalized Ito rule, 41

H
Hurwitz, 58, 65, 66

I
Increasing process, 22
Increment process, 18, 41
Indicator function, 50
Innovation process, 17
Invertible matrix, 38
Ito, 3, 17, 26, 41, 42, 44, 58

J
Jensen inequality, 21

K
Kalman filter, 11, 14

L
Lagrangian equations, 40, 43, 44
Lebesgue measure, 45
Lebesgue null set, 11, 13, 14, 22, 23, 46
Limit distribution, 19
Limit set, 4, 5, 11, 14–27, 36
Linear, time-invariant (LTI) systems, 3
Linear quadratic (LQ) performance, 3, 34–36
Lipschitz, 45
Locally input-to-state stable, 48, 65
Log-likelihood gradient, 12, 22, 23, 25

M
Martingale, 12, 17, 19, 21, 22, 25, 45, 54, 55, 58, 59, 63
Matrix eigenvalues, 4
Maximum likelihood (ML), 4, 9, 11–13, 34, 43

N
Non-singular matrix, 13

O
Optimal LQ performance, 3, 36

Ordinary differential equation (ODE), 43, 44, 47, 48, 52, 53, 61, 62, 64, 65

P
Persistent excitation (PE), 2, 4, 16, 21
Perturbed linear ODE, 48
Picard iterations, 57, 63–64
Positive-definite matrix, 66
Progressively measurable, 13
Projection, 5, 31, 36, 46, 51
Pseudo Hessian, 41

Q
Quadratic form, 50
Quadratic variation, 5, 43, 50, 58

R
Random limit, 14
Recursive algorithms, 35, 40
Recursive least squares (RLS), 1, 4, 13, 14, 21, 22, 45, 46, 51, 55
Regret rate, 54–56

S
Sample path, 3, 7, 9, 12, 14
Semi-martingale, 14, 15, 41
Set of indistinguishable dynamics, 4
σ-fields, 7, 13
Slack variable process, 41
Smoothed projected gradient approximation, 44
Stabilizable pairs, 4, 8, 22, 23
Stable equilibrium, 48
State dynamics matrix, 15, 17
State transition matrix, 64, 67, 68
Stochastic adaptive control (SAC), 1–2
Stochastic differential equations (SDEs), 2, 5, 16, 17, 26, 34, 40–43, 45, 47, 49, 51, 52, 58
Stopping time, 19, 51, 66
Strong consistency, 2
Subsequence, 18, 20, 23, 69
Sup-norm, 57, 67, 70
System regression vector, 2

T
Tangent space, 31, 36
Tensor, 38

The projection of $\nabla J(\theta)$, 31, 36
Time-varying constrained optimization, 2, 41
True system parameter, 8
Truncation mechanism, 15

U
Unforced ODE, 48

Uniform boundedness, 45, 46, 49
Unique minimum of J, 28
Unique strong solution, 5, 44, 45, 49
Unit eigenvector, 31

Z
Zero-order approximation, 60–61